MATH
FOR WORKPLACE SUCCESS

GENERAL BUSINESS

by
LLOYD D. BROOKS, ED.D.
Memphis State University

Consulting Editor:
DAVID J. PUCEL, PH.D.
University of Minnesota

ABOUT THE AUTHORS

Lloyd D. Brooks is chair of the Department of Management Information Systems and Decision Sciences at Memphis State University. He received his undergraduate degree from Middle Tennessee State University and graduate degrees from the University of Tennessee. He has researched and published widely in business math and computing, including several textbook and software publications in these areas. He has served as president of the Southern Business Education Association and has served on the executive board of the National Business Education Association. In addition, he has been named Tennessee Business Teacher of the Year and National Data Educator of the Year.

David J. Pucel, Consulting Editor, is professor and head of the division of industrial education, Department of Vocational and Technical Education, University of Minnesota. His degrees are in industrial education from the University of Wisconsin-Stout (B.A.), vocational education from the University of Illinois (M.A.), and education from the University of Minnesota (Ph.D.). He has conducted research, presented, and published extensively on instructional design, teaching and evaluation in vocational education, and training in business and industry, and he is an authority on performance-based instruction.

Editor: Art Lyons
Writing Services: Melanie Ruda
Production: Wiest Publications Management, Inc.
Illustrator: Tom Lochray

Library of Congress Cataloging-in-Publication Data

Brooks, Lloyd D., 1942-
Math for workplace success : general business / by Lloyd Brooks : consulting editor, David Pucel.
p. 364 cm.
ISBN 1-56118-257-5
1. Business mathematics. 2. Business mathematics--Problems, exercises, etc. I. Pucel, David J. II. Title.
HF5691.B763 1991 91-471
650'.01'513--dc20 CIP

Tax tables in Lesson 12 were adapted from the Internal Revenue Service's *Circular E: Employer's Tax Guide* (Publication 15, January 1991).

© 1991. Paradigm Publishing International
7500 Flying Cloud Drive, Eden Prairie, MN 55344

All rights reserved. No part of this publication may be reproduced, stored in a retrieval system, or transmitted, in any form or by any means, electronic, mechanical, photocopying, recording, or otherwise, without the prior written permission of Paradigm Publishing International.

Printed in the United States

10 9 8 7 6 5 4 3 2 1

TABLE OF CONTENTS

To The Student

Office workers perform different job tasks from company to company. Although the job tasks may differ, they often have a common requirement: an understanding of math. Office workers rely on their math skills to do such things as

- complete order forms
- calculate pay
- analyze sales and expense summaries
- understand the company's financial reports
- monitor inventory

What will this book teach students?

The purpose of this book is to teach problem-solving strategies for work problems that require math. Students who use this book will

- practice problem-solving strategies
- gather information from a variety of forms
- find amounts by adding, subtracting, multiplying and dividing
- compare partial and total amounts
- verify information
- analyze and interpret calculations to present information
- sharpen math skills

What does it mean to be good at math?

Being good at math means being able to apply math to the world of work. Knowing *how* to add and subtract is important, but it is not enough. People who are good at math know *when* to subtract or divide. They know *which numbers to use* and *what order to put them in*. They understand what their calculations mean and use the results to make decisions. Being good at math takes the kind of practice this book will provide.

How will this book help students with their math?

This book will prepare students to apply their math skills by teaching them a problem-solving strategy. Each lesson presents a new set of work situations in which the strategy is demonstrated. Students are guided through a practice problem before trying several on their own. Each lesson concludes with a review of related math skills and a progress check.

The next section, "How Good Problem Solvers Think about Math," explains the problem-solving strategy in detail.

How Good Problem Solvers Think About Math

A work problem is different from traditional math problems. Here is a typical math problem:

$$\begin{array}{r} 81 \\ \times\ 29 \\ \hline \end{array}$$

Notice that the math problem is already set up. The numbers and operation symbol (+, −, ×, or ÷) are there on the page. All that is required is to calculate.

Compare the math problem above to the work problem below:

Work Problem

Marge is a secretary who takes a monthly count of the paper supply at the office. She wrote her counts for January and February on the inventory form below. In February, Marge's boss noticed that the office was over budget for copier paper. (*Over budget* means that they had spent more on copier paper than they had planned.) She asked Marge to find out how much copier paper was used in January.

SUPPLY INVENTORY		
Item	**Jan. 1**	**Feb. 1**
boxes of fax paper	16	12
boxes of copier paper	15	2
boxes of computer paper	18	14

Notice that a typical work problem requires more thinking than a traditional math problem. Marge must decide which numbers to use and where to find them. She must choose an operation and set up the problem. Only then is she ready to calculate.

Like all good problem solvers, Marge follows a *strategy*. This section explains that problem-solving strategy. Refer to this section as often as necessary while working through this book.

Overview

1. **DEFINE** the problem.

   This step gives direction to the problem-solving process. The two questions below help define the problem:

   - What is the expected outcome?
   - What is the purpose?

2. **PLAN** the solution.

   Good problem solvers make sense of the problem by pulling it apart. They decide how to get the expected outcome by answering these four questions:

   - What data are needed?
   - Where can the data be found?
   - What is already known?
   - Which operation should be used?

3. **SOLVE** the problem.

   In this step, the plan is carried out. The data are selected and math skills are used to find the solution to the problem.

4. **CHECK** the solutions.

   Make sure the calculated solution solves the work problem. Arriving at a number does not guarantee that the work problem has been solved. Problem solvers refer back to the purpose and expected outcome. To be sure they solved the work problem, they ask the following:

   - Was the defined purpose accomplished?
   - Is the solution to the work problem reasonable?

   If the answers to these two questions are *yes*, then the problem-solving process is completed. If the answers are *no*, then the problem solver must work through the steps again to find the mistake and solve the problem.

Making Decisions about the Problem

Each question in the strategy requires the problem solver to make a decision. Here is an explanation of the questions and the answers they require.

DEFINE the problem

- *What is the expected outcome?*

The expected outcome is the kind of information that is being requested. It could be, for example, a total, sum, difference, increase, decrease, part, percent, ratio, verification, or ranking.

- *What is the purpose?*

The purpose is the *reason* for solving the problem. There are four basic workplace purposes considered in this book. Each one requires a different calculation process.

1. To find an amount.

Amount means *how much* or *how many*. If the expected outcome is how many in all (total), how many more or how many fewer (difference), how many groups, or how many in each group, the purpose is to find an amount. Finding how many dollars are spent on equipment, how many employees work in branch offices, how much oil is left in the tank, or how many hours each employee should work are all examples of finding amounts.

2. To express a relationship.

A relationship is a type of *comparison* between two numbers. When the expected outcome is a fraction, a ratio, or a percent, the purpose is to find a relationship. Finding the percent of a budget that is spent on equipment, the proportion (ratio) in which to mix two colors of dye, or what part of a shelf may be used for a display are all examples of expressing a relationship.

3. To verify.

   To verify is to *check.* When the purpose is to verify, the answer will be *yes* or *no.* Checking to make sure that the boss's expense report is accurate or that the amounts on a purchase order have been added correctly are examples of verifying numbers.

4. To analyze and interpret.

   To analyze and interpret is to *give meaning* to the calculations. If the expected outcome is a ranking of data or a decision based on that ranking, then the purpose is to analyze and interpret. Ranking sales representatives based on the number of items they sold or trucks based on the number of miles they travelled are examples of interpreting numbers. Awarding sales bonuses or scheduling a truck for service are examples of decisions that are made on the basis of these data interpretations.

DEFINE the problem

- *What data are needed?*

Data needed to solve a problem are the *numbers* needed to do the calculations. To find out how much pay an employee will earn, for example, one needs to know the dollars the employee earns per hour and how many hours were worked. Dollars per hour and hours worked are the data needed.

- *Where can the data be found?*

Work documents such as summaries, payroll ledgers, and order forms are often sources for needed data. Other sources are your own calculations and experience.

- *What is already known?*

Decide what information will be used to solve the problem. The work document may or may not have the exact data that is needed, but it will provide a starting place. An employee's time card, for example, will show hours worked. Another document, such as a list of employee pay rates, might be needed to find additional information.

- *Which operation should be used?* Problem solving often requires one operation or a combination of operations. Here are the four basic math operations and a brief description of when to use them.

1. **Addition**—Add to find a total, a sum, or how many in all.
2. **Subtraction**—Subtract to find a difference, an increase, a decrease, how much or how many more or how much less or how many fewer.
3. **Multiplication**—Multiply to find the total number of items in several equal groups.
4. **Division**—Divide to determine the number of groups or the number in each group.

SOLVE the problem

- *Select the relevant data.* Choose the numbers needed for the calculation.

- *Set up the calculations.* Numbers must be placed in the correct order or position for the operation. For example, in subtraction, the larger number is usually placed on top. In division, the number being divided is placed under the division sign.

- *Do the calculations.* Add, subtract, multiply, or divide the numbers.

- *Check the accuracy of the answer.* Accuracy is important on the job, so checking calculations is a must. To check, use the operation which is opposite of the one in the calculation. For example, use addition to check subtraction; use multiplication to check division. To check the total of several numbers, add the numbers in the reverse order.

Note: Expressing relationships, verifying, and analyzing and interpreting all require additional steps as part of the calculation process. These will be explained in each of the units.

CHECK the solution

Make sure the calculation solves the work problem.

Do not confuse this CHECK with the accuracy check that is part of the calculation process. This check focuses on the work problem. To do this check, refer back to the two questions from the DEFINE step.

- *Was the defined purpose accomplished?*

  Was the question, such as "How much" or "How many", answered? The answer should be an amount, a numerical relationship, a verification, or an interpretation.

- *Is the solution to the work problem reasonable?*

  Does the answer make sense? Think about how the answer should compare in size to the original numbers. For example, if the expected outcome is a *total*, the answer should be *larger* than the numbers added. If the expected outcome is a *difference*, the answer should be *smaller* than the original large number. In addition, ask whether your solutions make sense given the work problem.

Review

Look at the problem-solving strategy once more:

1. DEFINE the problem.	What is the expected outcome? What is the purpose?
2. PLAN the solution.	What data are needed? Where can the data be found? What is already known? Which operations should be used?
3. SOLVE the problem.	Set up, do, and check the calculations.
4. CHECK the solutions. Make sure the calculation solves the work problem.	Was the defined purpose accomplished? Is the solution to the work problem reasonableable?

UNIT I

FINDING AMOUNTS

In this unit, students will solve problems and find amounts when using a chart, completing payroll statements, requesting supplies, and ordering merchandise. When the solution involves finding an amount, the procedure below should be followed during the SOLVE step:

1. Select the relevant data.
2. Set up the calculation.
3. Do the calculation.
4. Check the accuracy of the answer.

Amounts are found in different ways: by adding and subtracting whole numbers, by adding and subtracting decimals, by multiplying and dividing whole numbers, and by multiplying and dividing decimals.

City Office Supply

The setting for this unit is City Office Supply, a company that sells office supplies and furniture. The work problems are related to the office, the accounting department, and the shipping department. Some problems involve City Office Supply's customers and their offices.

Carmen Hernandez and Rashad Delaney work at City Office Supply. Carmen manages the store. She decides what merchandise to carry, and she is responsible for the success of the store. Carmen relies on Rashad Delaney, the office clerk, to carry out the details of running the office. Rashad helps Carmen collect data for her reports, files information, and keeps track of inventory. (*Inventory* means the supply of goods on hand.) Everyone in the company knows Evelyn Garvey, the bookkeeper, because she is responsible for the paychecks. Ed Smiley is the shipping clerk who keeps track of the merchandise sent to customers.

LESSON 1

Finding Amounts by Adding and Subtracting Whole Numbers

When an office worker uses math on the job, the first step is to *find an amount.* Finding amounts with addition and subtraction is one of the most common math tasks in an office. Addition is used to find totals, such as the number of items sold or the amount of money in the cash register. Subtraction is used to calculate deductions from paychecks or the number of items left on the shelf after a sale.

Notice Rashad Delaney's solution to the following work problem.

Work Problem

City Office Supply sells six models of desk chairs. Carmen Hernandez, the manager, noticed that many more customers were buying chairs in February than in January. At the end of February, Carmen said to Rashad, "Please find out how many more chairs were sold this month than last month." Then she handed him the chart shown below. The chart was a record of how many chairs of each model were sold during January and February.

Companies often use charts like this one to summarize their sales of particular items. In this chart, the *rows* of numbers (numbers going across) show how many chairs were sold each month. The *columns* of numbers (numbers going down) show how many chairs of each model were sold. For example, City Office Supply sold nine (9) Executive-1 chairs in January and 12 in February.

CHAIR SALES

	MODEL						
MONTH	**Tech-1**	**Tech-2**	**Exec-1**	**Exec-2**	**Vis-1**	**Vis-2**	**TOTALS**
January	11	12	9	17	13	14	
February	16	32	12	18	11	12	
March							
April							
May							
June							

Rashad added all the numbers in the rows marked "January" and "February" to find out how many chairs were sold in each of those months. Add the numbers below and write the answer in the space provided.

Chairs sold in January	*Chairs sold in February*
11	16
12	32
9	12
17	18
13	11
+ 14	+ 12
______________ *chairs*	______________ *chairs*

Rashad found that City Office Supply sold 76 chairs in January and 101 chairs in February. Then he subtracted to find out how many more chairs were sold in February than in January, or the *difference*. Subtract the numbers below and write the answer in the space provided.

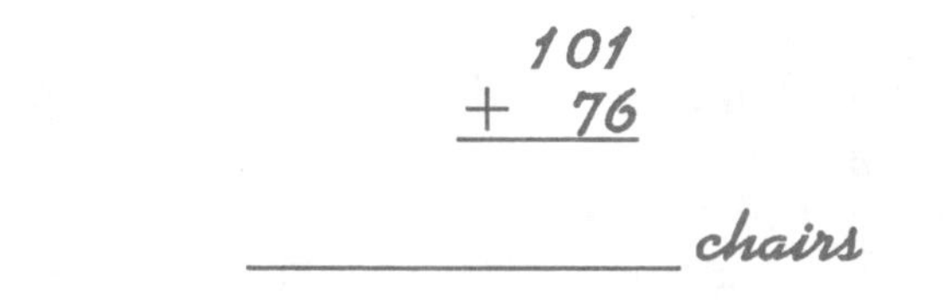

Rashad found that City Office Supply sold 25 more chairs in February than in January. Students who did not get the answers 101, 76 and 25 might need more practice in adding and subtracting. The Skills Practice section on page 17 reviews addition and subtraction for those who have trouble adding and subtracting in this lesson.

Doing Math to Find Amounts

How did Rashad know he needed to add and subtract? He defined the problem; he knew he was looking for an amount. Then he planned his solution.

1. He wanted to know the total number of chairs sold each month.

2. He looked on the chart, which listed the numbers he needed to find the totals.

3. The chart did not tell him the total number of chairs sold, but it did tell him how many chairs of each model were sold.

4. Rashad planned to *add* to find the total number of chairs sold and then *subtract* to find out how many more chairs were sold in February than in January.

Next, Rashad solved the problem. He followed four steps. These are the four steps that should be followed whenever an amount is needed.

1. **Select the relevant data.**

Select only the data that are needed to solve the problem. Remembering the expected outcome will help determine the data to select. Rashad selected the rows of data for January and February.

2. **Set up the calculations.**

To find the totals for January and February, Rashad added the numbers in the rows. Adding is usually easier if it is set up in a column.

January	February
11	16
12	32
9	12
17	18
13	11
+ 14	+ 12

3. **Do the calculations.**

Add the numbers in the columns.

January	February
76	101

4. **Check the accuracy of your answer.**

Rashad checked his work by adding in the opposite direction.

January	February
14	12
13	11
17	18
9	12
12	32
+ 11	+ 16
76	101

Sometimes, the data needed to solve a problem must be found by calculating. This may require working through the SOLVE steps a second time. The first time, the "given" data are used. The second time, the new, "calculated," data are used.

Rashad's problem is an example of this. In order to find out how many more chairs were sold in February, Rashad needed to know how many chairs were sold in January and in February so he could make a comparison. But Rashad did not know those numbers.

He did know how many chairs of *each model* were sold. He added those numbers to find out how many chairs were sold *each month*: 76 in January and 101 in February. Then he worked through the SOLVE steps again, as shown below, to find out how many more were sold in February than in January.

1. Select the data.

This time, the data are the new, calculated, totals for January and February.

2. Set up the calculation.

In his plan, Rashad said he would subtract. He knew that he should subtract because he wanted to find how many more.

$$\begin{array}{r} 101 \\ -\ 76 \\ \hline \end{array}$$

3. Do the calculation.

25

4. Check the accuracy of the answer.

Check subtraction by doing the opposite operation: addition. The answer should be the larger number you started with in the subtraction problem.

$$\begin{array}{r} 76 \\ +\ 25 \\ \hline 101 \end{array}$$

Rashad has his solution. In February, 25 more chairs were sold than in January.

Now an entire problem will be completed.

Work Problem

The two Visitor chairs are very similar in style, so Carmen is thinking about carrying only one of them next year. She says to Rashad, "I'd like to know how much difference there was in the sales of the two Visitor chairs last year. Can you use the information in this chart and give me the answer tomorrow morning?"

CHAIR SALES							
	MODEL						
MONTH	Tech-1	Tech-2	Exec-1	Exec-2	Vis-1	Vis-2	TOTALS
January	11	12	9	17	13	14	
February	16	32	12	18	21	12	
March	14	11	18	24	15	13	
April	15	20	16	13	18	17	
May	19	18	14	21	19	10	
June	10	17	22	21	11	12	
July	13	10	10	16	14	13	
August	17	13	11	15	25	16	
September	18	14	13	19	20	14	
October	10	15	15	10	22	11	
November	11	16	17	14	16	10	
December	12	7	9	11	10	9	
TOTALS							

DEFINE the problem

Rashad defined the problem by asking two questions.

- What is the expected outcome? *A difference*

- What is the purpose? *To find an amount*

NOTE: The word *difference* indicates that an *amount* is needed.

PLAN the solution

Once Rashad knew the purpose, he used these four questions to form the plan.

- What data are needed? *The yearly totals for models DC - 05 and DC - 06*

- Where can the data be found? *On the sales chart*

- What is already known? *The monthly totals for models DC - 05 and DC - 06*

- Which operations should be used? *Addition and subtraction*

SOLVE the problem

Rashad added and subtracted. He followed the four steps to SOLVE the problem.

- Select the relevant data. *Monthly totals for models DC - 05 and DC - 06*

- Set up the calculation.

Visitor-1	Visitor-2
13	14
21	12
15	13
18	17
19	10
11	12
14	13
25	16
20	14
22	11
6	10
+10	+ 9

- Do the calculations.

Add the monthly totals.

204	151

- Check the accuracy of the answer.

Add the numbers in reverse order.

Visitor-1	Visitor-2
10	9
16	10
22	11
20	14
25	16
14	13
11	12
19	10
18	17
15	13
21	12
+13	+14
204	151

Then Rashad did another calculation to find the difference in sales of the two chairs.

- Select the relevant data. *Monthly totals for the Visitor chair models*

- Set up the calculation.

$$\begin{array}{r} 204 \\ -151 \\ \hline \end{array}$$

- Do the calculation. 53

- Check the accuracy of the answer.

$$\begin{array}{r} 151 \\ +\ 53 \\ \hline 204 \end{array}$$

CHECK the solution

Make sure the calculated solution solves the work problem.

- Was the defined purpose accomplished? *Yes, 53 is the difference in sales of DC - 05 and DC - 06.*

- Is the solution to the problem reasonable? *Yes, 53 is smaller than 204.*

Note: When adding, the answer should be larger than the original amounts. When subtracting, the answer should be smaller than the original amounts. In this case, Rashad was subtracting.

Problem-Solving Practice

Use the DEFINE, PLAN, SOLVE, and CHECK steps to solve the work problem below.

Work Problem

Carmen is working on the annual sales report. She comes to Rashad and says, "Please find out how many chairs were sold last year. Here is the sales information."

CHAIR SALES							
	MODEL						
MONTH	Tech-1	Tech-2	Exec-1	Exec-2	Vis-1	Vis-2	TOTALS
January	11	12	9	17	13	14	
February	16	32	12	18	21	12	
March	14	11	18	24	15	13	
April	15	20	16	13	18	17	
May	19	18	14	21	19	10	
June	10	17	22	21	11	12	
July	13	10	10	16	14	13	
August	17	13	11	15	25	16	
September	18	14	13	19	20	14	
October	10	15	15	10	22	11	
November	11	16	17	14	16	10	
December	12	7	9	11	10	9	

DEFINE the problem

Think about what Carmen is asking Rashad to do. Write the answers to the questions on the lines provided.

- What is the expected outcome? ______________________

____

____

- What is the purpose?

PLAN the solution

Think about the numbers needed to solve the work problem. Write the answers to the questions on the lines provided.

- What data are needed?

- Where can the data be found?

- What is already known?

- Which operations should be used?

SOLVE the problem

- Select the relevant data.

- Set up the calculations.

- Do the calculations.

- Check the accuracy of the answers.

Work through the SOLVE steps again if necessary.

CHECK the solution

Make sure the calculation solves the work problem.

- Was the defined purpose accomplished? ______________________________

- Is the solution to the work problem reasonable? ______________________________

Answers to Problem-Solving Practice questions begin on page 29.

On Your Own

Here are some more work problems. Remember to DEFINE the problem, PLAN the solution, SOLVE the problem, and CHECK the solution to make sure it solves the work problem. Write the answers on the sales chart provided for each problem unless asked to do otherwise.

Work Problem A

On April 1, Carmen begins writing the quarterly report. She says, "There is a chart in the files about our desk sales. Please find out how many desks were sold during the first quarter of the year so it can be included in the report." The chart shown below gives sales numbers for each model for the first quarter.

SALES OF DESK MODELS							
	MODEL						
QUARTER	**Student-1**	**Student-2**	**Exec-1**	**Exec-2**	**Child-1**	**Child-2**	**TOTALS**
1st Quarter	18	19	22	18	20	23	
2nd Quarter							
3rd Quarter							
4th Quarter							
TOTALS							

Instead of dividing a year into twelve months, businesses often divide a year into four parts, or quarters. There are three months in each quarter.

First Quarter	Second Quarter	Third Quarter	Fourth Quarter
January	April	July	October
February	May	August	November
March	June	September	December

Businesses often keep track of their sales by the quarter, not by the month. Reports which summarize information for the quarter are called *quarterly* reports.

Work Problem B

Carmen would like to compare sales of the different desk models to find out which desks sold the best. She says, "Find out how many desks were sold of each model." Use the data in the chart shown below to get the information.

SALES OF DESK MODELS							
	MODEL						
QUARTER	**Student-1**	**Student-2**	**Exec-1**	**Exec-2**	**Child-1**	**Child-2**	**TOTALS**
1st Quarter	18	19	22	18	20	23	
2nd Quarter	20	15	19	14	18	17	
3rd Quarter	21	17	18	20	23	21	
4th Quarter	16	14	13	12	15	11	
TOTALS							

Work Problem C

It is time to prepare the annual report. Carmen says, "Find out how many desks were sold for the whole year." Use the data in the chart shown below to get the information.

SALES OF DESK MODELS							
	MODEL						
QUARTER	Student-1	Student-2	Exec-1	Exec-2	Child-1	Child-2	TOTALS
1st Quarter	18	19	22	18	20	23	
2nd Quarter	20	15	19	14	18	17	
3rd Quarter	21	17	18	20	23	21	
4th Quarter	16	14	13	12	15	11	
TOTALS							

Work Problem D

Carmen knows that City Office Supply sales decrease during the fourth quarter, but she wants to know how much they drop. She says, "Find the difference in sales between the third quarter and the fourth." Use the data in the chart shown below to get the information.

SALES OF DESK MODELS							
	MODEL						
QUARTER	Student-1	Student-2	Exec-1	Exec-2	Child-1	Child-2	TOTALS
1st Quarter	18	19	22	18	20	23	
2nd Quarter	20	15	19	14	18	17	
3rd Quarter	21	17	18	20	23	21	
4th Quarter	16	14	13	12	15	11	
TOTALS							

Work Problem E

Carmen is preparing a report to show the increase or decrease in the sales of various office machines. She says, "I need to know how much our sales decreased for typewriters and adding machines from Year 1 to Year 2. I also need to know how much our sales of calculators increased." Use the data in the chart shown below to get the information.

SALES OF OFFICE MACHINES 1989-1990			
	TYPEWRITERS	ADDING MACHINES	CALCULATORS
Year 1	245	173	448
Year 2	198	146	631

Answer: ____________________

Decrease in typewriter sales ____________________

Decrease in adding machine sales ____________________

Increase in calculator sales ____________________

Skills Practice: Adding and Subtracting Whole Numbers

UNDERSTANDING PLACE VALUE

What is similar about these three numbers?

321 132 213

All three of these numbers use the same digits, and all three numbers are 3-digit numbers. Although the digits are the same, each number has a different value because of the position, or *place value*, of the digits. *Number place* determines a number's actual value. Look at these three numbers when they are put on place value charts.

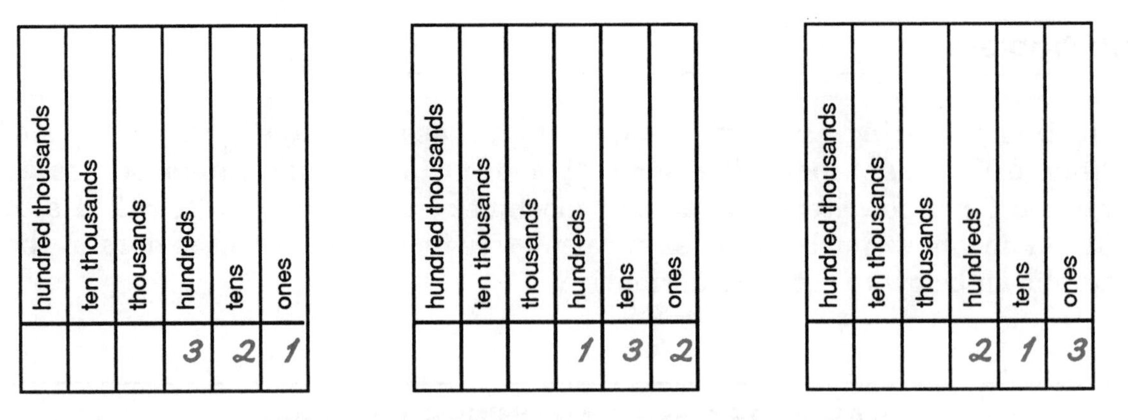

hundred thousands	ten thousands	thousands	hundreds	tens	ones
			3	2	1

hundred thousands	ten thousands	thousands	hundreds	tens	ones
			1	3	2

hundred thousands	ten thousands	thousands	hundreds	tens	ones
			2	1	3

The value of the position becomes higher as the number placement moves to the left. So, a 3 in the hundreds place has greater value than a 3 in the tens place. A 3 in the hundreds place stands for 3 hundreds, or 300. A 3 in the tens place stands for 3 tens, or 30.

Because different places have different values, digits must be lined up carefully to add and subtract. The ones should be put under the ones, the tens under the tens, hundreds under the hundreds, and so forth. A digit put in the wrong place changes the value of the number.

UNDERSTANDING ADDITION

Addition is the process of finding the total number of items in an entire group. Adding is a fast way of counting. For example, if there are 8 women and 5 men in an office, they could be counted to find out how many people there are altogether.

Or, the number of persons in each group could be *added*: $8 + 5$

Either way shows that there are 13 people in the office.

When is it appropriate to add? Often, adding is needed in response to requests like "find the sum," "find the total," or "find out how many in all."

Exercise 1

Write the phrases below in math form. Do *not* solve.

Example: the total of twelve and nine ______ $12 + 9$

1. the sum of fifty-five and eighty ______

2. nineteen more than thirty ______

3. the total of twenty-one and seventeen ______________

4. thirteen plus ninety-six ______________

5. the total of fifteen and eight ______________

Answers begin on page 31.

ADDITION WITH WHOLE NUMBERS

Step 1. When adding, line up numbers at the right, so that the digits in the ones place are aligned, like this:

$$\begin{array}{r} 53 \\ +\ 26 \\ \hline \end{array}$$

Step 2. Add the digits in the ones column.

$$\begin{array}{r} 53 \\ +\ 26 \\ \hline 9 \end{array}$$

Add 3 + 6.
Write the 9.

Step 3. Continue adding, one column at a time, moving to the left.

$$\begin{array}{r} 53 \\ +\ 26 \\ \hline 79 \end{array}$$

Add 5 + 2.
Write down the 7.

79 is the total, or *sum*.

Exercise 2

Add the following numbers. Write the answers below each problem.

$$\begin{array}{r} 23 \\ +\ 12 \\ \hline \end{array} \quad \begin{array}{r} 31 \\ +\ 23 \\ \hline \end{array} \quad \begin{array}{r} 14 \\ +\ 24 \\ \hline \end{array} \quad \begin{array}{r} 35 \\ +\ 21 \\ \hline \end{array} \quad \begin{array}{r} 21 \\ +\ 63 \\ \hline \end{array} \quad \begin{array}{r} 24 \\ +\ 21 \\ \hline \end{array}$$

Answers begin on page 31.

ADDITION WITH CARRYING

When adding, if the total for a column of digits is ten or more, carrying is needed. Follow this example.

Step 1. Add the digits in the ones column and carry any tens to the tens column.

```
   1
  798
+ 494   Add 8 + 4.
    2   Write down the 2.
        Carry the 1 to the top of the tens column because the 1 in the number 12 stands for 1 ten.
```

Step 2. Add the digits in the tens column and carry any hundreds to the hundreds column.

```
  11
  798
+ 494   Add 1 + 9 + 9.
   92   Write down the 9.
        Carry the 1 to the top of the hundreds column.
```

Step 3. Add the digits in the hundreds column.

```
  11
  798
+ 494   Add 1 + 7 + 4.
 1292   Write down the 12.
```

1292 is the sum.

Exercise 3

Add the following numbers. Write down the answers below each problem.

1.	45 + 67	28 + 32	33 + 67	47 + 56	59 + 25	43 + 64

2.

104	302	317	543	638	538
+ 423	+ 251	+ 816	+ 647	+ 475	+ 538

Answers begin on page 31.

Exercise 4

Add the following numbers. Write the answers below each problem.

1.

84	80	29	13	83	28
+ 53	+ 55	+ 59	+ 24	+ 58	+ 16

2.

109	353	274	485	373	364
+ 325	+ 465	+ 264	+ 153	+ 858	+ 936

3.

536	364	905	374	643	342
+ 428	+ 783	+ 852	+ 752	+ 638	+ 856

UNDERSTANDING SUBTRACTION

Subtraction is the process of finding the difference between two numbers. To subtract, take the the smaller value away from the larger value. There are two kinds of situations that require subtraction:

1. Subtract—take one amount from another amount—to find out *how much is left.* For example, if a store had nine computers and then six were sold, take 6 from 9 to find out how many are left.

$$\begin{array}{r} 9 \\ -\ 6 \\ \hline 3 \end{array}$$ computers are left

2. Subtract to find a *difference.* For example, if a store has seven large computers and two small computers, the difference between the number of large computers and small computers is five. In other words, there are five more large computers than small computers.

$$\begin{array}{r} 7 \\ -\ 2 \\ \hline 5 \end{array}$$ The answer, 5, is the difference between 7 and 2.

Subtraction is used to find a difference, an increase, a decrease, how many more, how many fewer, or how much is left.

The order of the numbers is important in subtraction. The large number usually is written first. The number being taken away (subtracted) is written second. If someone says, "Take 9 from 12," the larger number, 12, is written first and the number being taken away, 9, is written second. "Take 9 from 12" is written 12 − 9.

Exercise 5

Write the phrases below in math form. Do *not* solve.

Example: take two from eleven ___11 − 2___

1. subtract two from five ________
2. take nine from fourteen ________
3. three less than eight ________
4. deduct ten from twenty-two ________
5. the difference between seven and four ________
6. thirty-one minus twenty ________
7. six fewer than forty-three ________

Answers begin on page 31.

SUBTRACTION WITH WHOLE NUMBERS

As with addition, subtraction takes place digit by digit, beginning with the column on the right.

Step 1. Subtract the digits in the ones column.

$$\begin{array}{r} 48 \\ -\ 32 \\ \hline 6 \end{array}$$ Subtract, 8 − 2.
Write down the 6.

Step 2. Subtract the digits in the tens column.

```
  48
- 32   Subtract, 4 − 3.
  16   Write down the 1.
```

The difference is 16.

Exercise 6

Solve the following subtraction problems. Write the answers below each problem.

1.

```
  28      37      58      63      76
- 13    - 24    - 25    - 31    - 44
```

2.

```
  175     258     376     464     573
-  22   - 125   - 160   - 243   - 451
```

Answers begin on page 31.

SUBTRACTION WITH BORROWING

If a digit in the top number is smaller than the digit directly below it, borrow from the digit to the immediate left of the number on top. In the problem below, the 5 is smaller than the 8, so borrowing is necessary.

```
  35
- 18
```

Step 1. Borrow a ten from the 3, leaving 2 in the tens place. Show the borrowed 10 by crossing out the 5 and writing 15. Subtract the ones column.

```
   15   Cross out the 3 and write a 2.
  2̶3̶5̶   Cross out the 5 and write 15.
 - 18   Subtract, 15 − 8.
    7
```

Step 2. Subtract the tens column.

```
 2 15
  3 5
- 1 8   Subtract, 2 − 1.
  1 7
```

Exercise 7

Solve the following subtraction problems. Write the answers below each problem.

1.

```
  35      42      53      64      73
- 17    -  8    - 25    - 19    - 37
```

2.

```
  352     532     642     675     793
- 125   - 114   - 236   - 138   - 155
```

Answers begin on page 31.

Sometimes it is necessary to borrow more than once, as in this problem:

```
  735
- 259
```

Step 1. Borrow a 10 from the 3, leaving 2 tens. Show the borrowed 10 by crossing out the 5 and writing 15. Subtract the ones column.

```
   2 15
 7 3 5
-2 5 9   Subtract, 15 − 9.
     6
```

Step 2. Borrow a hundred from the 7, leaving 6 hundreds. Show the borrowed 100 by crossing out the 2 and writing 12. Subtract the tens column.

```
   6 12 15
    7  3  5
  − 2  5  9    Subtract, 12 − 5.
       7  6
```

Step 3. Subtract the hundreds column.

```
   6 12 15
    7  3  5
  − 2  5  9    Subtract, 6 − 2.
    4  7  6
```

Exercise 8

Solve the following subtraction problems. Write the answers below each problem.

1.

852	917	532	427	138
− 264	− 138	− 367	− 159	− 455

2.

732	614	735	773	920
− 349	− 546	− 469	− 575	− 273

SUBTRACTION FROM ZEROS

Sometimes it is necessary to subtract from a number that has one or more zeros. Look at the following example.

```
  600
− 132
```

Since two cannot be subtracted from zero, borrowing is required. However, there is nothing to borrow in the tens place. Therefore, borrow from the hundreds place.

Step 1. Borrow one hundred from the 6 (leaving 5). Show the borrowed 100 by crossing out the zero (in the tens place) and writing 10.

$$\begin{array}{r} {}^{5}\not{6}\ \overset{10}{\not{0}}\ 0 \\ -\ 1\ 3\ 2 \\ \hline \end{array}$$

Step 2. Borrow a ten from the 10, leaving 9. Show the borrowed 10 by crossing out the zero and writing 10.

$$\begin{array}{r} {}^{5}\not{6}\ \overset{\overset{9}{\not{10}}}{\not{0}}\ \overset{10}{\not{0}} \\ -\ 1\ 3\ 2 \\ \hline \end{array}$$

Step 3. Now the problem is ready for subtraction.

Exercise 9

Solve the following subtraction problems. Write the answers below each problem.

1.

$$\begin{array}{r} 200 \\ -\ \ 39 \\ \hline \end{array} \qquad \begin{array}{r} 500 \\ -\ \ 82 \\ \hline \end{array} \qquad \begin{array}{r} 300 \\ -\ \ 53 \\ \hline \end{array}$$

2.

$$\begin{array}{r} 302 \\ -\ 148 \\ \hline \end{array} \qquad \begin{array}{r} 407 \\ -\ 259 \\ \hline \end{array} \qquad \begin{array}{r} 803 \\ -\ 674 \\ \hline \end{array}$$

3.

$$\begin{array}{r} 6012 \\ -\ 2138 \\ \hline \end{array} \qquad \begin{array}{r} 2003 \\ -\ \ 785 \\ \hline \end{array} \qquad \begin{array}{r} 9005 \\ -\ 4317 \\ \hline \end{array}$$

Answers begin on page 31.

CHECKING SUBTRACTION

Check subtraction by adding the difference (the answer) to the second (the smaller) number. The result should equal the first (the larger) number in the original problem.

Example: **Subtraction Problem** **Answer Check**

$$\begin{array}{r} 891 \\ -\,342 \\ \hline 549 \end{array} \qquad \begin{array}{r} 342 \\ +\,549 \\ \hline 891 \end{array}$$

Exercise 10

Subtract the following and check your answer to the right of each problem.

1. $\begin{array}{r} 28 \\ -\,19 \\ \hline \end{array}$ $\begin{array}{r} 87 \\ -\,29 \\ \hline \end{array}$ $\begin{array}{r} 92 \\ -\,27 \\ \hline \end{array}$

2. $\begin{array}{r} 256 \\ -\,125 \\ \hline \end{array}$ $\begin{array}{r} 752 \\ -\,176 \\ \hline \end{array}$ $\begin{array}{r} 804 \\ -\,156 \\ \hline \end{array}$

3. $\begin{array}{r} 352 \\ -\,147 \\ \hline \end{array}$ $\begin{array}{r} 631 \\ -\,274 \\ \hline \end{array}$ $\begin{array}{r} 760 \\ -\,179 \\ \hline \end{array}$

Check Yourself

Circle the correct letter.

1. You should add when you hear the phrase

a. find the total.

b. find the difference.

c. find how many fewer.

d. find how many are left.

2. You should subtract when you hear the phrase

a. find the total.

b. find the difference.

c. find the sum.

d. find how many altogether.

3. In which work situation would you NOT be finding an amount:

a. when totaling the amount in the cash register.

b. when copying numbers from the payroll ledger to the payroll statement.

c. when finding out how many book shelves are left.

d. when finding out how many tables have been sold.

4. The strategy steps occur in this order:

a. PLAN, DEFINE, SOLVE, CHECK.

b. PLAN, SOLVE, DEFINE, CHECK.

c. DEFINE, PLAN, SOLVE, CHECK.

d. DEFINE, PLAN, CHECK, SOLVE.

5. How many months are there in one quarter of a year?

a. 1/4

b. 3

c. 6

d. 4

Work Problem

City Office Supply sells 2-drawer, 3-drawer and 4-drawer filing cabinets. Carmen has decided to make a graph showing how sales decreased during the year for various kinds of merchandise, including filing cabinets. She says, "Find out how much the sales of filing cabinets decreased between the first and second quarters." Use the following chart to solve the problem and put your numbers in the chart.

SALES OF FILING CABINETS				
	MODEL			
QUARTER	2-Drawer	3-Drawer	4-Drawer	TOTALS
1st Quarter	47	44	51	
2nd Quarter	40	39	48	
3rd Quarter	41	38	43	
4th Quarter	36	33	35	
TOTALS				

Answers to Problem-Solving Practice Questions

DEFINE the problem

- To find how many chairs were sold (an amount)
- A total, "how many in all"

PLAN the solution

- How many chairs of each model were sold during the year
- From the monthly totals for each chair listed on the chart
- How many chairs of each model were sold each month
- Adding

SOLVE the problem

- The monthly totals for each of the 4 chair models for which no year's totals are given.

Technical-1	Technical-2	Executive-1	Executive-2
11	12	9	17
16	32	12	18
14	11	18	24
15	20	16	13
19	18	14	21
10	17	22	21
13	10	10	16
17	13	11	15
18	14	13	19
10	15	15	10
11	16	17	14
12	7	9	11
166	185	166	199

• 12	7	9	11
11	16	17	14
10	15	15	10
18	14	13	19
17	13	11	15
13	10	10	16
10	17	22	21
19	18	14	21
15	20	16	13
14	11	18	24
16	32	12	18
11	12	9	17
166	185	166	199

Not all the necessary information was obtained by adding together the monthly totals. The steps must be repeated.

- The number of chairs of each model sold during the year
- 166
  185
  166
  199
  204
  151
- 1071
- 151
  204
  199
  166
  185
  166

  1071

CHECK the solution

- The purpose for calculating was accomplished. The amount, 1071, represents the number of chairs sold during the year.
- The result of the calculations, 1071, does provide a reasonable solution to the work problem because it is a number larger than the original amounts.

Answers to Skills Practice Problems

Exercise 1

55 + 80	19 + 30	21 + 17	13 + 96	15 + 8

Exercise 2

35	54	38	56	84	45

Exercise 3

1.	112	60	100	103	84	107
2.	527	553	1133	1190	1113	1076

Exercise 5

1. 5 – 2 **2.** 14 – 9 **3.** 8 – 3 **4.** 22 – 10 **5.** 7 – 4 **6.** 31 – 20 **7.** 43 – 6

Exercise 6

1.	15	13	33	32	32
2.	153	133	216	221	122

Exercise 7

1.	18	34	28	45	36
2.	227	418	406	537	638

Exercise 9

1.	161	418	247
2.	154	148	129
3.	3874	1218	4688

LESSON 2

Finding Amounts by Adding and Subtracting Decimals

Businesses, like people, must keep track of the money they earn and spend. Keeping track of money requires an understanding of how to find amounts by adding and subtracting *decimals*. Addition and subtraction of decimals is also necessary to find some amounts, such as the price of gasoline, the wage rate for employees, amounts on a utility bill, or charges for a long distance telephone call.

Notice Evelyn Garvey's solution to the following work problem.

Work Problem

Evelyn Garvey is the bookkeeper at City Office Supply. She completes *payroll statements* for each employee. A payroll statement lists an employee's *earnings* and *deductions* for each pay period. Earnings include regular pay and, possibly, a bonus or overtime pay. Earnings are often called *gross pay*. Deductions are amounts taken out of pay by the employer, such as taxes and insurance. Deductions are subtracted from the gross pay to find the *net pay*.

Gross pay – deductions = net pay

Net pay is the amount of a paycheck and is sometimes called *take home pay*.

Today is October 6. The end of the pay period is near, and Evelyn must calculate Rashad's net pay so she can prepare his check.

CITY OFFICE SUPPLY

PAYROLL STATEMENT

Employee *Rashad Delaney*

Employee No. *301* Department *560*

Payroll Classification *9-2* S. Sec. No. ______

PAYROLL PERIOD From *September 25, 199–*

To *October 8, 199–*

EARNINGS			DEDUCTIONS		
Salary	653	60	F.I.C.A.	43	55
Overtime	16	34	Fed'l W.H. Tax	73	70
Tips			Unemployment Ins.	33	50
			State W.H. Tax	26	80
Total Earnings			Total Deductions		
Net Pay					

To solve this problem, Evelyn added Rashad's earnings to compute total earnings (or gross pay) and then added his deductions to compute total deductions. Add and write your totals below:

Earnings	*Deductions*
	$43.55
	73.70
$653.60	33.50
+ 16.34	+ 26.80
Total Earnings $	*Total Deductions $*

Evelyn found Rashad's total earnings to be $669.94 and his total deductions to be $177.55. Then Evelyn subtracted the total deductions from the total earnings to get his net pay. Do the subtraction and write the difference below the line.

$669.94	(Total Earnings)
− 177.55	(Total Deductions)
	(Net Pay)

Rashad's net pay is $492.39. Students whose answers are different from Evelyn's might need more practice in adding and subtracting decimals. The section on page 48 provides additional assistance and practice adding and subtracting decimals.

Doing Math to Find Amounts

Evelyn's purpose is to find an amount, Rashad's net pay. Here is what she knows from the plan she made.

1. Evelyn must know total earnings and total deductions.
2. She can find total earnings and total deductions by adding the amounts on the payroll statement.
3. She knows salary, overtime, FICA, federal withholding tax, unemployment insurance, and state withholding tax.
4. First, she will add all the earnings together. Then, she will add all the deductions together. Finally, she will subtract the total deductions from the total earnings.

Next, Evelyn solved the problem. When she was calculating, Evelyn followed four steps. These are the same four steps to be followed when solving a problem to get an amount.

1. Select the relevant data.	*Rashad's earnings and deductions*	
2. Set up the calculations.	*Earnings* $ 653.60 + 16.34	*Deductions* $ 43.55 73.70 33.50 + 26.80
3. Do the calculations.	$ 669.94	$ 177.55
4. Check the accuracy of your answers.	$ 16.34 +653.60 $ 669.94	$ 26.80 33.50 73.70 + 43.55 $ 177.55

Evelyn now has the data she needs to calculate Rashad's net pay. Once again, she works through the steps needed to calculate an amount.

1. Select the relevant data.	*Rashad's total earnings and total deductions*	
2. Set up the calculation.	$ 669.94 − 177.55	*Total Earnings* *Total Deductions*
3. Do the calculation.	$ 492.39	*Net Pay*
4. Check the accuracy of the answer.	$ 177.55 +492.39 669.94	

Now, an entire problem will be completed.

Work Problem

On October 21, at the end of the pay period, Evelyn must prepare the paychecks. She must calculate Susan Kim's net pay on the payroll statement shown below.

CITY OFFICE SUPPLY

PAYROLL STATEMENT

Employee *Susan Kim*

Employee No. *304* Department *560*

Payroll Classification *S-3* S. Sec. No. ______

PAYROLL PERIOD From *October 9, 199–* To *October 22, 199–*

EARNINGS			DEDUCTIONS		
Salary	281	00	F.I.C.A.	18	26
Overtime			Fed'l W.H. Tax	30	91
Tips			Unemployment Ins.	14	05
			State W.H. Tax	11	24
Total Earnings			Total Deductions		
Net Pay					

DEFINE the problem

- What is the expected outcome? *Net pay*

- What is the purpose? *To find an amount*

PLAN the solution

- What data do I need? *Total earnings and total deductions*

- Where can I find it? *On the payroll statement*

- What do I already know? *Total earning and four deductions: F.I.C.A, federal withholding tax, unemployment insurance and state withholding tax*

- Which operations should I use? *Addition and subtraction*

SOLVE the problem

The payroll slip shows that the only earnings are $281.00, or the salary. Therefore, the total earnings equal the salary. Evelyn needed to find the total deductions, so she added the individual deductions.

- Select the relevant data. *Earnings and deductions*

- Set up the calculation.

Deductions

$18.26
30.91
14.05
+ 11.24

- Do the calculation. $74.46

- Check the accuracy of the answer.

Deductions

$11.24
14.05
30.91
+ 18.26
$74.46

Evelyn needed to do another calculation to finish the problem.

- Select the relevant data. *Total earnings and total deductions*

- Set up the calculation.

$281.00
− 74.46

- Do the calculation. $206.54

- Check the accuracy of the answer.

$74.46
+ 206.54
$281.00

CHECK the solution

Make sure the calculated solution solves the work problem.

- Was the defined purpose accomplished? *Yes, $206.54 is Susan's net pay.*

- Is the solution to the work problem reasonable? *Yes, $206.54 is smaller than $281.00.*

Evelyn then filled in the amounts for total earnings, the total deductions, and the net pay on the payroll statement.

Problem-Solving Practice

Use the DEFINE, PLAN, SOLVE, and CHECK steps to solve the work problem below.

Work Problem

At City Office Supply, part of the responsibility of the bookkeeper is to complete the payroll statements once every two weeks. The end of the pay period is approaching. Complete the following payroll statement for Frank Cordoza.

CITY OFFICE SUPPLY

PAYROLL STATEMENT

Employee *Frank Cordoza*

Employee No. *306* Department *560*

Payroll Classification *C-1* S. Sec. No. ______

PAYROLL PERIOD 〉 From *October 23, 199–* To *November 5, 199–*

EARNINGS			DEDUCTIONS		
Salary	972	80	F.I.C.A.	71	14
Overtime	121	60	Fed'l W.H. Tax	107	01
Tips			Unemployment Ins.	54	72
			State W.H. Tax	43	78
Total Earnings			Total Deductions		
Net Pay					

DEFINE the problem

- What is the expected outcome? ______

- What is the purpose? ______

PLAN the solution

Think about the numbers needed to solve the problem.

- What data are needed? ____________________
- Where can the data be found? ____________________
- What is already known? ____________________
- Which operations should be used? ____________________

SOLVE the problem

- Select the relevant data. ____________________
- Set up the calculations.
- Do the calculations.

- Check the accuracy of the answers.

Work through the SOLVE steps again if necessary.

CHECK the solution

Make sure the calculated solution solves the work problem.

- Was the defined purpose accomplished? ______________________________
____
____

- Is the solution to the work problem reasonable?

Answers to Problem-Solving Practice questions begin on page 57.

On Your Own

Here are some more work problems. Remember to DEFINE the problem, PLAN the solution, SOLVE the problem, and CHECK the solution to be sure it solves the work problem.

Work Problem A

The bookkeeper at City Office Supply calculates the net pay for all employees and completes the payroll statements. Find Rashad's net pay for the pay period of November 6 to November 19 and complete the payroll statement below.

CITY OFFICE SUPPLY

PAYROLL STATEMENT

Employee Rashad Delaney

Employee No. 303 **Department** 560

Payroll Classification M-1 **S. Sec. No.**

PAYROLL PERIOD **From** November 6, 199–
To November 19, 199–

EARNINGS			DEDUCTIONS		
Salary	976	80	F.I.C.A.	69	84
Overtime	97	68	Fed'l W.H. Tax	118	19
Tips			Unemployment Ins.	53	72
			State W.H. Tax	42	98
Total Earnings			Total Deductions		
Net Pay					

Work Problem B

Jeff Josephs earns $9.08 an hour as a full-time bookkeeper. From November 20 to December 3, Jeff worked his usual 40 hours per week (for a total salary of $726.40) plus an extra 6 hours (for a total overtime pay of $54.48). Complete Jeff's payroll statement and find his net pay.

CITY OFFICE SUPPLY

PAYROLL STATEMENT

Employee *Jeff Josephs*

Employee No. *307* Department *560*

Payroll Classification *B-1* S. Sec. No.

PAYROLL PERIOD From *November 20, 199–*

To *December 3, 199–*

EARNINGS			DEDUCTIONS		
Salary			F.I.C.A.	50	76
Overtime			Fed'l W.H. Tax	85	90
Tips			Unemployment Ins.	39	04
			State W.H. Tax	31	24
Total Earnings			Total Deductions		
Net Pay					

Work Problem C

It is now two weeks later. Jeff has not worked any overtime from December 4 to December 17. How much less will Jeff bring home for this pay period than last?

Answer: ____________________

CITY OFFICE SUPPLY

PAYROLL STATEMENT

Employee *Jeff Josephs*

Employee No. *307* Department *560*

Payroll Classification *B-1* S. Sec. No. ________

PAYROLL PERIOD

From *December 4, 199–*

To *December 17, 199–*

EARNINGS			DEDUCTIONS		
Salary	726	40	F.I.C.A.	47	22
Overtime			Fed'l W.H. Tax	79	90
Tips			Unemployment Ins.	36	32
			State W.H. Tax	29	06
Total Earnings			Total Deductions		
Net Pay					

Work Problem D

On December 18, Jeff got a raise. He now makes $9.28 per hour. From December 18 to December 31, he worked 40 hours a week with no overtime and earned $742.40. How much has Jeff's gross pay increased? How much has his net pay increased?

Answers:

Gross pay increase ____________

Net pay increase ____________

CITY OFFICE SUPPLY

PAYROLL STATEMENT

Employee *Jeff Josephs*

Employee No. *307* Department *560*

Payroll Classification *B-2* S. Sec. No. ____________

PAYROLL PERIOD From *December 18, 199–* To *December 31, 199–*

EARNINGS			DEDUCTIONS		
Salary	742	40	F.I.C.A.	48	26
Overtime			Fed'l W.H. Tax	81	66
Tips			Unemployment Ins.	37	12
			State W.H. Tax	29	70
Total Earnings			Total Deductions		
Net Pay					

Work Problem E

On January 1, state tax rates were increased. This means the deductions increased. Find Jeff's new net pay for the pay period of January 1 to January 14. Complete the payroll statement.

CITY OFFICE SUPPLY

PAYROLL STATEMENT

Employee *Jeff Josephs*

Employee No. *307* Department *560*

Payroll Classification *B-2* S. Sec. No. ______

PAYROLL PERIOD ⟩ From *January 1, 199–* To *January 14, 199–*

EARNINGS			DEDUCTIONS		
Salary	742	40	F.I.C.A.	48	26
Overtime			Fed'l W.H. Tax	81	66
Tips			Unemployment Ins.	37	12
			State W.H. Tax	31	55
Total Earnings			Total Deductions		
Net Pay					

Skills Practice: Adding and Subtracting Decimals

UNDERSTANDING DECIMALS

Lesson 1 presented adding and subtracting whole numbers. What happens when adding or subtracting an amount less than one? An *amount less than one* is a *decimal*. A common use of decimals is to display money. For example, $.43 is an amount less than one dollar; $.43 is a decimal. The decimal $.43 is shown on the following place value chart.

hundred thousands	ten thousands	thousands	hundreds	tens	ones	decimal point	tenths	hundredths	thousandths
						.	4	3	

Places to the left of the decimal point are for numbers equal to or greater than one. Places to the right of the decimal point are for numbers less than one. The decimal $.43 is less than $1. The farther a number is to the right of the decimal point, the less its value is. For example, a number in the hundredths position is smaller than a number in the tenths position.

Numbers are read according to their position in the place value chart. When reading numbers, the word *and* is used for the decimal point, as follows.

Number	**Read**
5.3	"five and three tenths"
18.25	"eighteen and twenty-five hundredths"
1.318	"one and three hundred eighteen thousandths"

Exercise 1

Write the following numbers.

1. ______________ seven tenths

2. ______________ fourteen hundredths

3. ______________ five hundredths

4. ______________ two hundred thirteen thousandths

5. ______________ thirty-five thousandths

6. ______________ four and two tenths

7. ______________ thirty seven and twenty-eight hundredths

8. ______________ two hundred five and four hundred thirty-one thousandths

Answers begin on page 58.

Exercise 2

Rewrite each of the amounts below in order from smallest to largest.

1.	2.9	29	0.29	______	______	______
2.	0.9	9	90	______	______	______
3.	14	14.1	1.42	______	______	______
4.	$23.47	$3.98	$23.00	______	______	______
5.	0.079	0.79	7.9	______	______	______

Answers begin on page 59.

ADDING DECIMALS

Addition of decimal values is similar to addition of whole numbers. A special word of caution: the decimal point of each number must be lined up with the decimal point of the number above it. Follow this example.

The decimals below are to be added.

30.4 + 23.708 + 7.18

Step 1. Write the numbers so that each decimal point is placed directly below the decimal point in the preceding number.

```
   30.4
   23.708
+   7.18
```

Step 2. Find the total. Notice that all decimal points are aligned. Remember to carry if necessary.

```
   30.4
   23.708
+   7.18
   61.288
```

Exercise 3

Add.

1.

2.3	4.5	6.8	4.3	5.7	7.8
3.48	5.321	4.503	5.327	2.74	5.29
3.27	4.28	3.342	4.382	2.871	3.24

2.

14.2	13.5	24.534	31.4	42.371
3.08	25.783	31.56	4.28	5.392
2.784	7.203	30.083	25.324	72.3

Answers begin on page 59.

Exercise 4

Rewrite the numbers below in columns, making sure the decimal points are aligned. Then add.

1. 3.5 + 2.81 + 5.206 =

2. 14.3 + 28.56 + 1.9 =

3. 247.1 + 2.8 + 3.807 =

4. 472.3 + 728.3 + 0.017 =

5. 682.3 + 0.057 + 4.03 =

6. 4.0187 + 4.3 + 6.004 =

7. 0.03702 + 5.3 + 53.901 =

8. 6.7 + 0.0365 + 3.2341 =

9. 35.2 + 4.004 + 7.873 =

10. 8.2301 + 4.03 + 9.7 =

SUBTRACTING DECIMALS

The same process for subtracting whole numbers can be used for subtracting decimals. However, as with adding decimals, be sure the decimal points in all the numbers line up, as shown below:

$$\begin{array}{r} 389.58 \\ \underline{-\,125.42} \\ 264.16 \end{array}$$

Sometimes a subtraction problem looks like the one below.

325.4
− 18.36

What is the 6 subtracted from? Place a zero to the right of the 4, as shown below:

325.40
− 18.36

Zeros may be written to the right of the last digit after the decimal point without changing the value of the number. Look at 325.4 and 325.40 on place value charts.

hundred thousands	ten thousands	thousands	hundreds	tens	ones	decimal point	tenths	hundredths	thousandths
			3	2	5	.	4		

hundred thousands	ten thousands	thousands	hundreds	tens	ones	decimal point	tenths	hundredths	thousandths
			3	2	5	.	4	0	

Both charts show 3 hundreds, 2 tens, 5 ones, 4 tenths and no hundredths. There are no hundredths in the first chart because the space is blank. There are no hundredths in the second chart because there is a zero in the hundredths place. Putting a 0 after the 4 simply emphasizes that there are no hundredths. The value has not been changed.

After a zero has been placed to the right of the 4, carry out the subtraction. If borrowing is necessary, follow the same process used when subtracting whole numbers.

325.40
− 18.36
307.04

Exercise 5

Subtract.

1. 34.43 − 21.22   47.83 − 43.71   53.72 − 14.51   64.19 − 27.02   78.40 − 35.11

2.	126.43 – 75.12	143.81 – 82.86	728.47 – 172.23	835.02 – 509.20	915.37 – 713.23
3.	17.35 – 15.3	27.48 – 13.2	64.28 – 39.5	83.53 – 52.8	93.71 – 24.5

Answers begin on page 59.

Exercise 6

Subtract.

1.	1.78 – 0.562	2.8 – 1.752	12.7 – 9.52	87.51 – 17.3582	58.5 – 14.78
2.	548.096 – 132.4	508.538 – 89.5	689.7 – 123.825	638.5 – 409.256	786.3 – 372.187

For decimals, as with whole numbers, a difference can be checked by adding, as shown below:

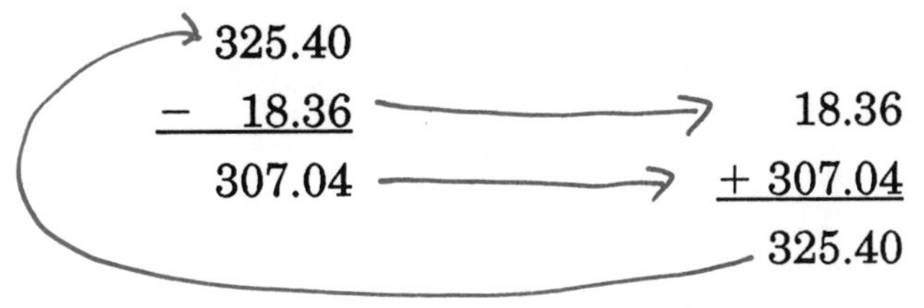

Exercise 7

Subtract. Then check each answer.

1. 47.85 − 13.14 = Check:

2. 89.34 − 23.21 = Check:

3. 1.489 − 0.231 = Check:

4. 2.536 − 1.245 = Check:

5. 8.525 − 3.5 = Check:

6. 9.782 − 5.8 = Check:

7. 7.354 − 2.93 = Check:

8. 74.28 − 19.3 = Check:

9. 83.3 − 2.464 = Check:

10. 487.5 − 9.362 = Check:

Check Yourself

1. Which of these situations would involve a decimal amount?
  a. counting boxes on the shelf
  b. calculating take home pay
  c. totaling the charges on a bill
  d. b and c
  e. a, b, and c

2. Gross pay is the same as ________________
  a. take home pay.
  b. total earnings.
  c. total deductions.
  d. net pay.

3. Which of the following is NOT a deduction?
  a. state tax
  b. F.I.C.A.
  c. unemployment insurance
  d. overtime pay

4. To calculate net pay, ________________________
  a. subtract total deductions from gross pay.
  b. subtract gross pay from total deductions.
  c. add gross pay to the deductions.
  d. add deductions to total earnings.

5. The first step when calculating is to ________________
  a. set up the calculation.
  b. do the calculation.
  c. select the relevant data.
  d. check the accuracy of the answer.

Work Problem

On January 26, Frank Cordoza says, "Don't forget that the pay period is almost over and the payroll statements are due by the end of the day." Complete the payroll statement below.

CITY OFFICE SUPPLY

PAYROLL STATEMENT

Employee *Evelyn Garvey*

Employee No. *302* Department *560*

Payroll Classification *B-2* S. Sec. No. ______

PAYROLL PERIOD 〉 From *January 15, 199–*

To *January 28, 199–*

EARNINGS			DEDUCTIONS		
Salary	738	40	F.I.C.A.	54	00
Overtime	92	30	Fed'l W.H. Tax	91	38
Tips			Unemployment Ins.	41	54
			State W.H. Tax	35	30
Total Earnings			Total Deductions		
Net Pay					

Answers to Problem-Solving Practice Questions

DEFINE the problem

- a difference
- to find an amount, net pay

PLAN the solution

- total earnings and total deductions
- from calculations with the earnings and deductions on the payroll statement
- For earnings: salary and overtime. For deductions: F.I.C.A., federal withholding tax, state withholding tax and unemployment insurance
- addition and subtraction

SOLVE the problem

- earnings and deductions

CITY OFFICE SUPPLY

PAYROLL STATEMENT

Employee *Frank Cordoza*

Employee No. *306* Department *560*

Payroll Classification *C-1* S. Sec. No. ______

PAYROLL PERIOD From *October 23, 199—* To *November 5, 199—*

EARNINGS			DEDUCTIONS		
Salary	972	80	F.I.C.A.	71	14
Overtime	121	60	Fed'l W.H. Tax	107	01
Tips			Unemployment Ins.	54	72
			State W.H. Tax	43	78
Total Earnings	1094	40	Total Deductions	276	65
Net Pay	817	75			

- **to find total earnings:**

```
 $ 972.80
+  121.60
 $1094.40
```

- **to check total earnings:**

```
 $1094.40
−  121.60
 $ 972.80
```

- **to find total deductions:**

```
 $ 43.78
   54.72
  107.01
+  71.14
 $276.65
```

- **to find net pay:**

```
 $1094.40
−  276.65
 $ 817.75
```

- **to check net pay:**

```
 $ 276.65
+  817.75
 $1094.40
```

CHECK the solution

- **The defined purpose was accomplished because $817.75 is an amount which represents Frank's net pay.**
- **The solution to the work problem, $817.75, is reasonable because it is smaller than $1094.40.**

Answers to Skills Practice Problems

Exercise 1

1. .7
2. .14
3. .05
4. .213
5. .035
6. 4.2
7. 37.28
8. 205.431

Exercise 2

1.	0.29	2.9	29
2.	0.9	9	90
3.	1.42	14	14.1
4.	$3.98	$23.00	$23.47
5.	0.079	0.79	7.9

Exercise 3

1.	9.05	14.101	14.645	14.009	11.311	16.33
2.	20.064	46.486	86.177	61.004	120.063	

Exercise 5

1.	13.21	4.12	39.21	37.17	43.29
2.	51.31	60.95	556.24	325.82	202.14
3.	2.05	14.28	24.78	30.73	69.21

LESSON 3

Finding Amounts by Multiplying and Dividing Whole Numbers

Office workers must find total amounts when taking inventory or completing requisitions. Adding is one way to find a total; *multiplying* is another. Totals found through multiplication are called *products*. Multiplying is much quicker than adding when there are several equal groups to total.

Sometimes the total is known, but the number of groups or the number in each group is not known. In these cases, the operation to use is *division*. Packers use division to decide how many cartons are needed or how many items should go in each carton. Purchasing clerks often use division to find the *unit price* of an item, or "how much each one costs."

Companies often have a room or area where they keep office supplies, such as paper and pencils. When employees need supplies, they complete a *requisition* (or order) form and give it to the person in charge of supplies. An order for three boxes of paper clips might look like the one below.

CITY OFFICE SUPPLY REQUISITION

DEPARTMENT__________ NAME______________________ DATE______________

ITEM #	QUANTITY	UNIT	DESCRIPTION
PC-013	3	boxes	paper clips

Notice the word *boxes* under the heading UNIT. A unit refers to how the item is counted and ordered. In this example, the paper clips are ordered by the box, not individually or by the carton. Items may be counted individually, by the dozen, the "gross" (twelve dozen), the package, the pound, the foot, or one of many other units of measure.

Notice Ed Smiley's solution to the following work problem.

Work Problem

Ed works in the shipping department, where they ran out of tape to seal the boxes. Ed completed a requisition and sent it to the supply clerk. In shipping, Ed uses 24 rolls of tape a week, and he wants to order enough tape for four weeks. Rolls of tape come in packages of six.

CITY OFFICE SUPPLY REQUISITION

DEPARTMENT *shipping* NAME *Ed Smiley* DATE *5/4*

ITEM #	QUANTITY	UNIT	DESCRIPTION
ST-303		*package*	*3 inch wide strapping tape*

First Ed multiplied 4 (weeks) by 24 (rolls per week) to find out how many rolls to order. Multiply 4 × 24 and write the answer here:

____

Ed found that they needed 96 rolls. (Students who found a different answer might want to work on multiplying whole numbers. See page 72.) Then Ed divided 96 (rolls needed) by 6 (rolls per package) to find out how many packages to order. Divide 96 by 6 and write the answer on the *Quantity* line on the requisition above.

Students who did not write 16 on the line might need more practice with division of whole numbers. The section that begins on page 72 provides extra practice and assistance in multiplication and division.

Doing Math to Find Amounts

Ed DEFINED his problem as needing to find an amount, how many packages of tape to order. This was his PLAN:

1. He needed to know how many rolls of tape would be used in four weeks and how many rolls are in a package.
2. He needed to calculate how many rolls would be used. He already knew how many rolls of tape come in a package.
3. He knew how quickly the tape was used and the number of weeks for which he must order.
4. He planned to multiply to find how many rolls were needed and then to divide to find out how many packages to order.

Next, Ed proceeded with the SOLVE step of the strategy. He calculated how many rolls of tape would be used. He followed the procedure shown below.

1. Select the relevant data. — *Rate at which tape is used (24 rolls per week) and number of weeks (4)*
2. Set up the calculation. — 24×4
3. Do the calculation. — *96*
4. Check the accuracy of the answer. (Note: Ed checked his answer by dividing.) — $96 \div 4 = 24$

Ed did another calculation to determine how many packages of tape to order.

1. Select the relevant data. — *How may rolls of tape would be used (96) and number of rolls per package (6)*
2. Set up the calculation. — $6\overline{)96}$

3. Do the calculation.

$$\begin{array}{r} 16 \\ 6\overline{)96} \\ \underline{60} \\ 36 \\ \underline{36} \\ 0 \end{array}$$

4. Check the accuracy of the answer.
(Note: Ed checked his answer by multiplying)

$$\begin{array}{r} 16 \\ \underline{\times\ 6} \\ 96 \end{array}$$

Ed needed to order 16 packages.

Now an entire problem will be completed.

Work Problem

As the shipping clerk, one of Ed's responsibilities is to keep track of supplies and order more when something is needed. The shipping department would soon be out of mailing labels, so Ed ordered more. He wanted enough to last six weeks, and the department uses about 200 labels a week. The labels come in packages of 250.

DEFINE the problem

- What is the expected outcome? *How many packages*
- What is the purpose? *To find an amount*

PLAN the solution

- What data are needed? *The number of labels to be used and the number of labels per package*
- Where can the data be found? *I know it from experience*
- What is already known? *How many labels we use per week (200), how many weeks I need to order for (6) and how many labels come in packages (250)*
- Which operations should be used? *Multiplication and division*

SOLVE the problem

The first calculation will determine the number of labels needed for six weeks.

- Select the relevant data. *The number of labels we use per week (200) and the number of weeks I need to order for (6)*
- Set up the calculation. 200×6
- Do the calculation. *1200*
- Check the accuracy of the answer. $1200 \div 6 = 200$

Shipping will use 1200 labels in six weeks.

The second calculation will determine how many packages to order.

- Select the relevant data.

  Number of labels needed (1200) and number of labels in a package (250)

- Set up the calculation.

  $250\overline{)1200}$

- Do the calculation.

$$\begin{array}{r} 4 \text{ R}200 \\ 250\overline{)1200} \\ \underline{1000} \\ 200 \end{array}$$

- Check the accuracy of the answer.

$$\begin{array}{r} 250 \\ \times \quad 4 \\ \hline 1000 \\ + \quad 200 \\ \hline 1200 \end{array}$$

Note: Check the answer by multiplying and then adding the remainder.

CHECK the solution

Make sure the calculation solves the work problem.

- Was the defined purpose accomplished?

  No. 4 R200 does not solve the work problem. It isn't possible to order 4 R200 packages. So Ed ordered 5 packages.

Note: When dividing results in a remainder, decide what the remainder means. An understanding of this meaning is important to get a realistic solution to the work problem. In a situation like the one above, the remainder of 200 means that Ed would be 200 labels short if he ordered 4 packages. He ordered 5 packages so he would have enough.

- Is the solution to the work problem reasonable?

Yes, 5 is smaller than 1200. Also, Ed knows it is reasonable for the department to use 5 packages of labels in 6 weeks

Note: In multiplication of whole numbers, the answer should be larger than the numbers being multiplied. In division of whole numbers, the answer should be smaller than the number being divided. In this case, Ed was dividing.

Based on his calculations, Ed completed the requisition as shown below. Notice that 5 goes in the column labeled *Quantity.*

CITY OFFICE SUPPLY REQUISITION

DEPARTMENT *shipping* NAME *Ed Smiley* DATE *6/30*

ITEM #	QUANTITY	UNIT	DESCRIPTION
ML-325	*5*	*packages*	*mailing labels*

Problem-Solving Practice

Use the DEFINE, PLAN, SOLVE, and CHECK steps to solve the next problem.

Work Problem

Carl, who works in the accounting department, runs out of paper for the adding machine. He uses two rolls a week and he wants to order enough for eight weeks. Adding machine tape comes in packages of three rolls. Write the order on the following requisition.

CITY OFFICE SUPPLY REQUISITION

DEPARTMENT *accounting* NAME *Carl Miller* DATE *6/8*

ITEM #	QUANTITY	UNIT	DESCRIPTION
MT-725		*packages*	*adding machine tape*

DEFINE the problem

- What is the expected outcome? ______________________________

- What is the purpose? ______________________________

PLAN the solution

- What data are needed? ______________________________

- Where can the data be found? ______________________________

- What is already known? ______________________________
____
____

- Which operations should be used? ______________________________

SOLVE the problem

- Select the relevant data. ______________________________
____
____

- Set up the calculation.

- Do the calculation.

- Check the accuracy of the answer.

Work through the calculation steps again if necessary.

CHECK the solution

Make sure the calculated solution solves the work problem:

- Was the defined purpose accomplished? ______________________________
____
____

- Is the solution to the work problem reasonable? ______________________

Answers to Problem-Solving Practice questions appear on page 86.

On Your Own

Here are some more work problems to try. Remember to DEFINE the problem, PLAN the solution, SOLVE the problem, and CHECK the solution. Make sure the solution solves the work problem.

Work Problem A

Ken is filing some receipts in the office when he realizes there are only enough file folders to last three more days. He uses about 50 folders a week. The folders come in boxes of 25. Order enough boxes to last 12 weeks. Write the order on the requisition below.

CITY OFFICE SUPPLY REQUISITION

DEPARTMENT *office* NAME *Ken Ray* DATE *7/23*

ITEM #	QUANTITY	UNIT	DESCRIPTION
99-67		*boxes*	*file folders*

Work Problem B

Charlie Pope, the new shipping clerk, submits the following requisition. Charlie did not know that he should have ordered the markers and pens by the box instead of individually. Markers come in boxes of six and pens come in boxes of twelve. How many boxes of each should the supply clerk give to Charlie?

Answers: Black Markers ________ boxes

Red Pens ________ boxes

CITY OFFICE SUPPLY REQUISITION

DEPARTMENT *shipping* NAME *Charlie Pope* DATE *8/19*

ITEM #	QUANTITY	UNIT	DESCRIPTION
FM-555	*35*	*each*	*black markers*
BP-151	*40*	*each*	*red pens*

Work Problem C

Carmen says, "We are almost out of paper for the copier. Please order more." The office makes about 400 copies a week. One package of paper, called a *ream,* contains 500 sheets of paper. Order enough reams for eight weeks.

CITY OFFICE SUPPLY REQUISITION

DEPARTMENT *office* NAME *Carmen Quinn* DATE *8/31*

ITEM #	QUANTITY	UNIT	DESCRIPTION
CP-987		*reams*	*white copier paper*

Work Problem D

Maria is preparing payroll statements while working in the accounting department. She notices there is only one pad of statements left. She uses 32 payroll statements each pay period. A pad has 50 sheets and a package has 3 pads. Order enough packages to last ten pay periods. Write the order on the following requisition.

CITY OFFICE SUPPLY REQUISITION

DEPARTMENT *accounting* NAME *Maria Johns* DATE *9/13*

ITEM #	QUANTITY	UNIT	DESCRIPTION
PS-072		*packages*	*payroll statement pads*

Work Problem E

Frank Cordoza, the accountant, says, "We're running out of memo pads. Please see that we get some more." Evelyn Garvey, the bookkeeper, overhears this and adds, "We're almost out of staples, too. Please order some staples." The department uses three memo pads a week; the pads come in packages of five. Staples come in boxes of 5000, and the office uses a box every two weeks. Order enough memo pads and staples to last 12 weeks.

CITY OFFICE SUPPLY REQUISITION

DEPARTMENT *accounting* NAME *Frank Cordoza* DATE *10/3*

ITEM #	QUANTITY	UNIT	DESCRIPTION
MP-728		*packages*	*pink memo pads*
MS-645		*boxes*	*standard staples*

Skills Practice: Multiplying and Dividing Whole Numbers

UNDERSTANDING MULTIPLICATION

Multiplication is a short way of adding two or more numbers. Three rows of envelopes with five envelopes in each row may be totaled by adding 5 + 5 + 5 to get 15 or by multiplying 3 rows by 5 envelopes to get 15.

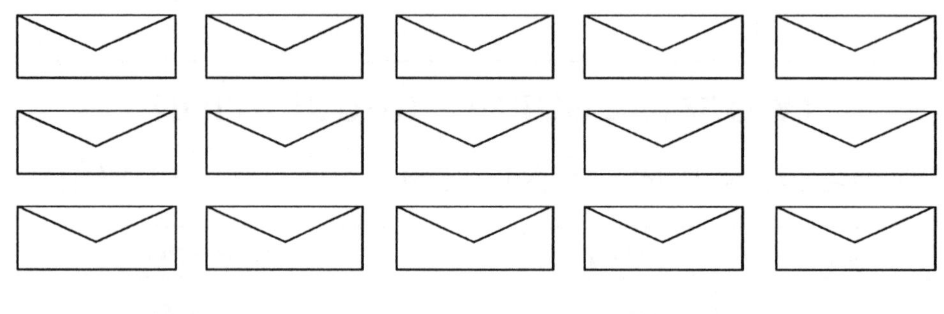

Addition	**Multiplication**
$5 + 5 + 5 = 15$	$3 \times 5 = 15$

When adding, the result is referred to as a sum, and when multiplying, the result is referred to as a product. To find a total or to find out "how many in all," either add or multiply. Add when the groups vary in amount. Multiply when a given number of groups has the same number in each group. In the example above, there were three groups with five envelopes in each group or five groups with three envelopes in each group.

Exercise 1

Circle the correct operation (add or multiply) in the situations below. Do NOT solve.

1. One box has 50 envelopes and another has 100. What is the total number of envelopes in the two boxes?

add multiply

2. Sandra Yee has 50 boxes of envelopes. Each box has 100 envelopes. How many envelopes does Sandra have in all?

add multiply

3. George Michaels has 18 boxes of pens with six pens in each box. What is the total number of pens?

add multiply

4. Andrew Johnson has three boxes of green pens. One box has six pens, one box has 12 pens, and one box has eight pens. How many pens does Andrew have in all?

add multiply

5. Gary Brown makes $5 an hour. He worked 20 hours. How much money did he earn?

add multiply

6. Gary earned $25 dollars yesterday. He earned $5 today. How much money did he earn?

add multiply

Answers begin on page 87.

MULTIPLICATION WITH WHOLE NUMBERS

To solve multiplication problems, students must memorize the multiplication facts or refer to a multiplication table like the one below. To use the table, find one of the numbers being multiplied (a *factor*) in the row across the top and the other factor in the column down the left side. The answer or product is found at the point where the row and the column meet. For example, the table tells you that $3 \times 3 = 9$.

MULTIPLICATION TABLE

	1	**2**	**3**	**4**	**5**	**6**	**7**	**8**	**9**	**10**	**11**	**12**
1	1	2	3	4	5	6	7	8	9	10	11	12
2	2	4	6	8	10	12	14	16	18	20	22	24
3	3	6	9	12	15	18	21	24	27	30	33	36
4	4	8	12	16	20	24	28	32	36	40	44	48
5	5	10	15	20	25	30	35	40	45	50	55	60
6	6	12	18	24	30	36	42	48	54	60	66	72
7	7	14	21	28	35	42	49	56	63	70	77	84
8	8	16	24	32	40	48	56	64	72	80	88	96
9	9	18	27	36	45	54	63	72	81	90	99	108
10	10	20	30	40	50	60	70	80	90	100	110	120
11	11	22	33	44	55	66	77	88	99	110	121	132
12	12	24	36	48	60	72	84	96	108	120	132	144

The following procedures show how to multiply. Begin with the digit in the ones place of the *bottom* number.

$$\begin{array}{r} 21 \\ \underline{\times\ 4} \end{array}$$ Begin with the digit in this place.

Step 1. Multiply the top, right-hand digit by the bottom digit in the ones place.

$$\begin{array}{r} 21 \\ \underline{\times\ 4} \\ 4 \end{array}$$

Multiply, 4×1.

Write the 4.

Step 2. Multiply the digit in the tens place by the bottom digit in the ones place.

$$\begin{array}{r} 21 \\ \underline{\times\ 4} \\ 84 \end{array}$$

Multiply, 4×2.

Write the 8.

The answer, 84, is called the *product*.

Exercise 2

Multiply.

1. $\begin{array}{r} 22 \\ \underline{\times\ 3} \end{array}$ $\begin{array}{r} 11 \\ \underline{\times\ 8} \end{array}$ $\begin{array}{r} 32 \\ \underline{\times\ 3} \end{array}$ $\begin{array}{r} 12 \\ \underline{\times\ 4} \end{array}$ $\begin{array}{r} 14 \\ \underline{\times\ 2} \end{array}$ $\begin{array}{r} 11 \\ \underline{\times\ 9} \end{array}$

2. $\begin{array}{r} 34 \\ \underline{\times\ 2} \end{array}$ $\begin{array}{r} 42 \\ \underline{\times\ 2} \end{array}$ $\begin{array}{r} 24 \\ \underline{\times\ 2} \end{array}$ $\begin{array}{r} 11 \\ \underline{\times\ 7} \end{array}$ $\begin{array}{r} 13 \\ \underline{\times\ 3} \end{array}$ $\begin{array}{r} 44 \\ \underline{\times\ 2} \end{array}$

Answers begin on page 87.

CARRYING WITH MULTIPLICATION

If the answer obtained by multiplying a particular digit is greater than nine, the number to the left of the first digit is *carried* .

Step 1. Multiply the digit in the ones place and carry.

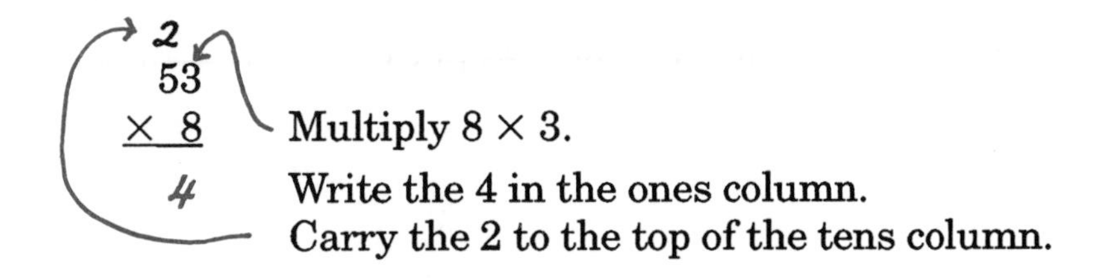

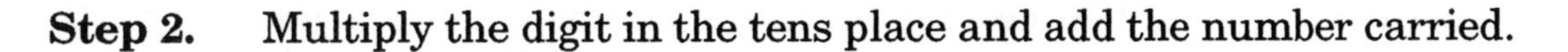

Step 2. Multiply the digit in the tens place and add the number carried.

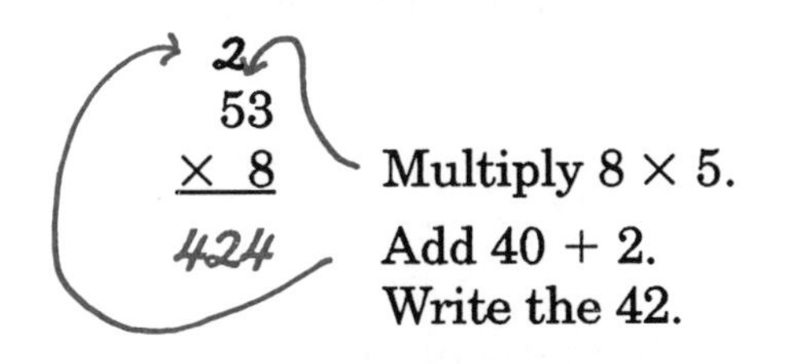

The product is 424.

Exercise 3

Multiply.

1. 24×8 25×7 36×5 59×9 244×7

2. 258×6 306×9 467×7 890×5 725×4

Answers begin on page 88.

MULTIPLYING BY TWO-DIGIT NUMBERS

A number is often multiplied by a two-digit number, as in 12 × 23. For this type of computation, the problem is split and the number is multiplied separately by each digit in the second number. Then those partial answers are combined.

Part 1. Multiply by the bottom digit in the ones place.

$$\begin{array}{r} 23 \\ \times\ 12 \\ \hline 46 \end{array}$$

Part 2. Multiply by the bottom digit in the tens place, again beginning on the right and moving to the left. IMPORTANT: When multiplying times a digit in the *tens* place, write the first digit of the answer in the *tens* place.

Step 1.

$$\begin{array}{r} 23 \\ \times\ 12 \\ \hline 46 \\ 3 \end{array}$$

Multiply 1 × 3.

Write the 3 in the tens place.

Step 2.

$$\begin{array}{r} 23 \\ \times\ 12 \\ \hline 46 \\ 23 \end{array}$$

Multiply 1 × 2.

Write the 2 to the left of the 3.

Part 3. Add each of the partial answers to get the product.

$$\begin{array}{r} 23 \\ \times\ 12 \\ \hline 46 \\ +23 \\ \hline 276 \end{array}$$

The product is 276.

Exercise 4

Multiply.

1.	20 ×12	30 ×14	50 ×23	60 ×35	70 ×26
2.	709 × 12	608 × 23	407 × 35	502 × 44	132 × 57

Answers begin on page 88.

Exercise 5

Multiply.

1.	1423 × 40	1524 × 70	3425 × 25	3645 × 36	2376 × 47
2.	1524 × 120	2636 × 350	4536 × 460	4732 × 142	5836 × 453

Exercise 6

Multiply.

1. 25 times 12 =

2. 43 times 23 =

3. 242 times 10 =

4. $25 \times 36 =$

5. $45 \times 128 =$

6. $52 \times 208 =$

7. $46 \times 309 =$

UNDERSTANDING DIVISION

Given a certain number of equal groups, multiply to find the total. Division is the opposite of multiplication. So, in division, start with the total and divide to find either the number of groups or the number in each group.

Here is an example. Evelyn has 18 envelopes. If she uses six each day, in how many days will all the envelopes be used?

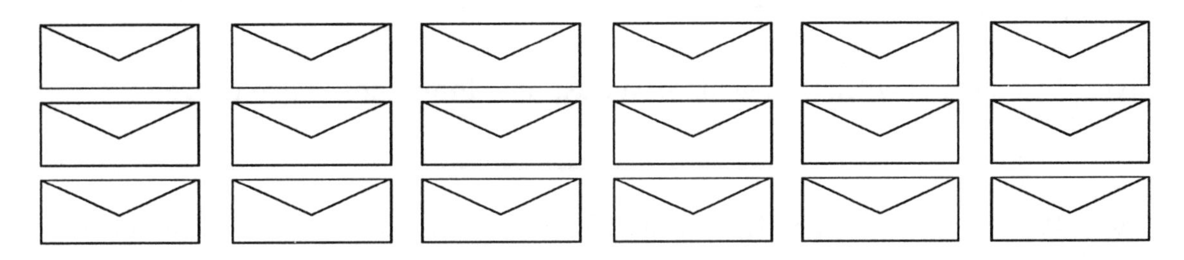

The total is 18 and the number in each group is 6. There are 3 groups of 6 envelopes, so it took 3 days to use the envelopes: $18 \div 6 = 3$.

Either of two signs may be used to show division: $\div$ or $\overline{)\quad}$.

When the sign $\div$ is used, the *total* is written first. If 24 is being divided into 4 groups, write $24 \div 4$. Read this as *24 divided by 4.*

When the sign $\overline{)\quad}$ is used, the total is written *underneath* or *inside* the division sign. If 24 is being divided into four groups, write

$$4\overline{)24}$$

Read this as *24 divided by 4* or *4 divided into 24.*

Divide when the solution requires the number in each group or the number of groups. If the word *each* or *per* is in the expected outcome, division is probably needed.

Exercise 7

Set up these problems. Use addition, subtraction, multiplication, or division. Do NOT solve.

Example: Given 100 rolls of tape with 4 rolls in a package, how many packages are there?

$100 \div 4$
____

1. Given 144 pencils with 12 pencils in a box, how many boxes are there?

____

2. Given 30 boxes of paper clips with 100 in a box, how many paper clips are there?

____

3. A store had 120 calendars; 60 were sold. How many were left?

____

4. Given 30 red pens and 40 blue pens, what is the total number of pens?

____

5. The supply shelf has 25 reams of white paper and 5 reams of yellow paper. How much more white paper is there than yellow paper?

____

6. A store served 100 customers a day for 5 days. How many customers were served in all?

____

7. Twenty boxes were used to pack 200 computer books. How many books were packed in each box?

____

8. Seventy calculators were packed. Five were put into each carton. How many cartons were used?

____

Answers begin on page 88.

DIVIDING WITH WHOLE NUMBERS

The division process requires four basic steps:

(1) divide, (2) multiply, (3) subtract, and (4) bring down the next digit. These steps are repeated until the problem is complete, as explained below. When dividing, the result is referred to as the *quotient*.

Assume that 152 must be divided by 2.

Divisor $2\overline{)152}$ Dividend

Step 1. **Divide** the *divisor* into the first digit of the *dividend*. If the first digit in the dividend is smaller (as in this problem), divide the divisor into the first *two* digits in the dividend.

$$\begin{array}{r} 7 \\ 2\overline{)152} \end{array}$$

Think: "Two cannot divide into one, so divide 2 into 15. 15 ÷ 2 = 7, with some left over."

Write the 7 on top, *above the 5*, because 5 is the right-hand digit of the number divided in this step (15).

Step 2. **Multiply** the divisor by the number obtained in the first step. Place the product directly below the digits used in the dividend.

$$\begin{array}{r} 7 \\ 2\overline{)152} \\ 14 \end{array}$$

Multiply 7× 2.
Write the product,14, under the 15.

Step 3. **Subtract** the product from the digits above it.

$$\begin{array}{r} 7 \\ 2\overline{)152} \\ \underline{14} \\ 1 \end{array}$$

Subtract 15 − 14.
Write the 1.

Step 4. **Bring down** the next digit of the dividend. Place it directly to the right of the number obtained in Step 3.

$$\begin{array}{r} 7 \\ 2\overline{)152} \\ \underline{14} \\ 12 \end{array}$$

Bring down the 2.
Place the 2 to the right of the 1.

Now the process of divide, multiply, subtract and bring down begins again with the number 12.

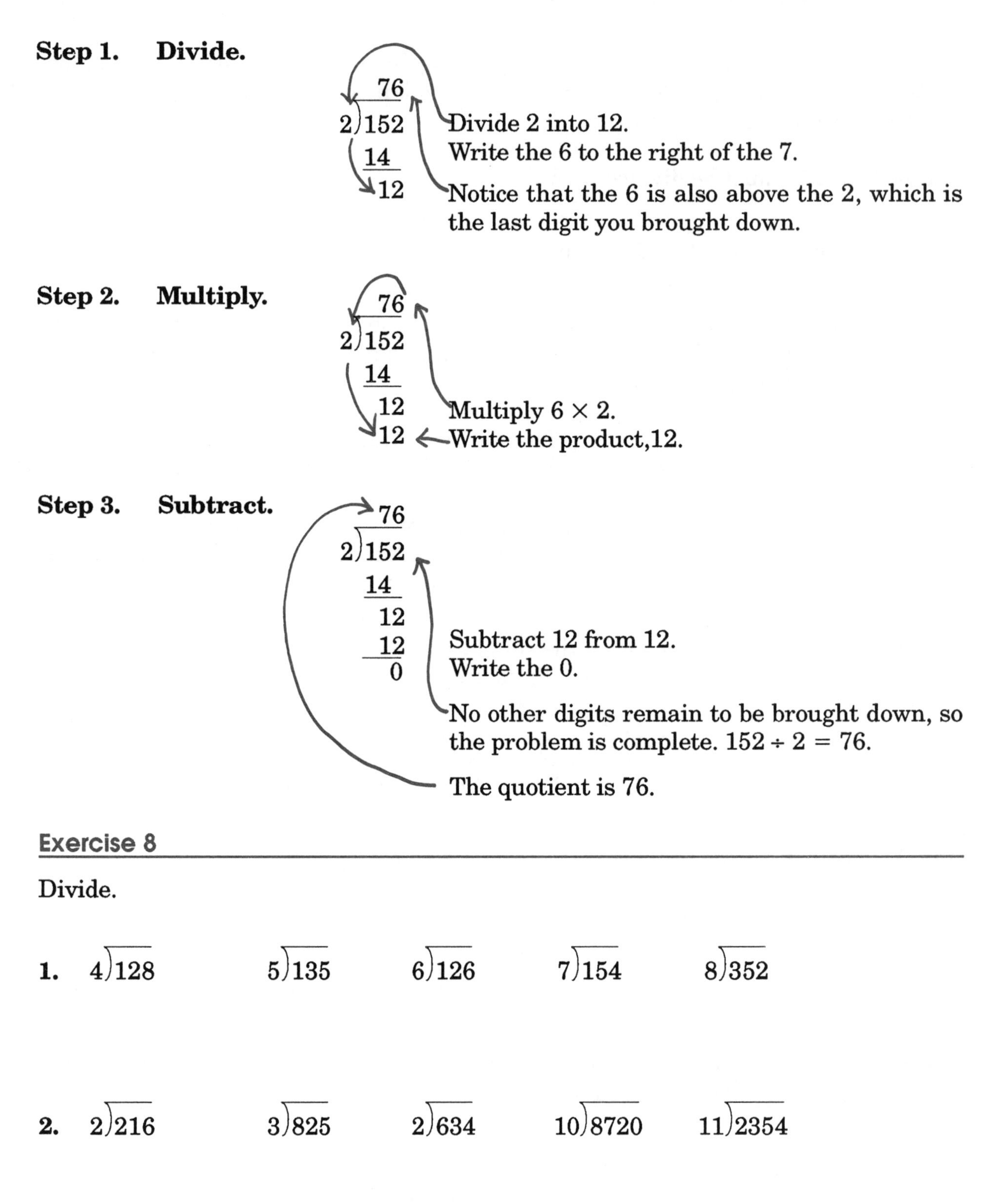

Answers begin on page 88.

DIVISION WITH REMAINDERS

Sometimes, the divisor does not divide into the dividend evenly and a remainder results. A slightly more complicated example will be used to show how remainders are handled. However, the same four basic steps will be repeated.

Assume that 257 is divided by 15.

```
     1    Step 1. Divide 15 into 25.
15)257    Step 2. Multiply 1 × 15.
   15     Step 3. Subtract 25 − 15.
   107    Step 4. Bring down the 7.
```

The above steps are repeated until all digits in the dividend are used, as shown below. If a number is left after completion of these steps, the number becomes the remainder.

```
    17   R 2   Write the remainder on top.
15)257
   15
   107
   105
     2   Remainder
```

If in one of the repetitions of the division steps, the divisor is larger than the number being divided, put a zero (0) in the answer and bring down the next digit in the dividend. In the example below, the second digit in the quotient is zero (0) because 8 cannot be divided into 2. Therefore, a zero is placed in the quotient and the next digit in the dividend is brought down.

```
   203    R 5
8)1629    Step 1. Bring down the 2.
  16      Step 2. Put a 0 on top
   029    because 8 cannot be
    24    divided into 2.
     5    Step 3. Bring down the 9.
          Continue as before.
```

Hint: Every time a digit is brought down, a digit must be placed in the quotient.

To check the answer, multiply the quotient by the divisor. Then, add the remainder. The result should equal the dividend. The check for the previous problem is shown below:

```
  203
×   8    Multiply 8 × 203.
 1624
+   5    Add the remainder.
 1629    The result, 1629, is the original dividend.
```

Exercise 9

Divide. Check the answers on a separate piece of paper.

1. $2\overline{)117}$ $5\overline{)127}$ $6\overline{)189}$ $8\overline{)194}$ $10\overline{)223}$

2. $11\overline{)312}$ $12\overline{)253}$ $13\overline{)300}$ $14\overline{)356}$ $15\overline{)698}$

Answers begin on page 88.

Exercise 10

Divide. Check the answers on a separate piece of paper.

1. $12\overline{)2772}$ $13\overline{)1768}$ $14\overline{)2156}$ $15\overline{)2745}$ $16\overline{)448}$

2. $8\overline{)1851}$ $6\overline{)1949}$ $5\overline{)1337}$ $10\overline{)8423}$ $9\overline{)920}$

3. $6\overline{)645}$ $8\overline{)847}$ $12\overline{)1250}$ $7\overline{)7289}$ $4\overline{)4276}$

Check Yourself

1. Circle the letter of the situation requiring multiplication.
   a. The same number of markers is in each of 36 boxes. Lloyd needs to know the total number of markers.
   b. A different number of markers is in each of 36 boxes and Lloyd needs to know the total number of markers.
   c. Lloyd wants to find out how many boxes to use if he has 36 markers and he puts the same number of markers in each box.
   d. Lloyd has 36 markers left after selling the same number and he wants to know how many he started with.

2. Circle the letter of the situation requiring division.
   a. Sharon has 36 boxes with the same number of markers in each box and she wants to find the total number of markers.
   b. Sharon has 36 boxes with a different number of markers in each box and she wants to find the total number of markers.
   c. Sharon wants to find out how many boxes to use if she has 36 markers and she puts the same number of markers in each box.
   d. Sharon has 36 markers left after selling the same number and she wants to know how many she started with.

3. Use a requisition ___________
   a. to find out what merchandise was sold to a customer.
   b. to order merchandise from another company.
   c. to order merchandise within the company.
   d. to keep track of how much merchandise is on the shelves.

4. Evelyn must order enough legal pads to last eight weeks. Her department uses five pads a week. Legal pads come in packages of three, and Evelyn must order by the package. What is Evelyn's expected outcome?
   a. Number of packages of legal pads
   b. Number of people in her department
   c. Number of weeks
   d. Number of legal pads

5. Charles Pope must order 50 bottles of glue. Glue comes in boxes containing 6 bottles. How many boxes should he order?
   a. 8 R 2
   b. 8
   c. 9
   d. 2

Work Problem

The account clerks at City Office Supply are responsible for mailing the billing statements. They use 350 envelopes a month. There are 100 envelopes in a box and 8 boxes in a carton. Order enough cartons of envelopes on the requisition below to last for six months.

CITY OFFICE SUPPLY REQUISITION

DEPARTMENT *accounting* NAME ____________ DATE *11/4*

ITEM #	QUANTITY	UNIT	DESCRIPTION
WE-711		*cartons*	*business envelopes*

Answers to Problem-Solving Practice Questions

DEFINE the problem

- How many rolls of tape, a total
- To find an amount

PLAN the solution

- Number of rolls needed and the number of rolls per package
- From the problem or situation
- Two rolls a week are used and the order is for eight weeks. A package has three rolls.
- Multiplication and division

SOLVE the problem

- Two rolls a week are used. The order is for eight weeks.
- $2 \times 8 =$
- 16
- $16 \div 8 = 2$
- To find how many are needed
- Sixteen rolls are needed; a package has three rolls.
- $16 \div 3 =$
- 5 R1
- $5 \times 3 + 1 = 16$

CHECK the solution

- The defined purpose was not accomplished. Carl cannot order 5 R1 packages. He must order 6 packages.
- The solution to the work problem, 6, is reasonable because it is less than 16.

CITY OFFICE SUPPLY REQUISITION

DEPARTMENT *accounting* NAME ____________ DATE *6/8*

ITEM #	QUANTITY	UNIT	DESCRIPTION
M7-725	*6*	*packages*	*adding machine tape*

Answers to Skills Practice Problems

Exercise 1

1. add **2.** multiply **3.** multiply **4.** add **5.** multiply **6.** add

Exercise 2

1.	66	88	96	48	28	99
2.	68	84	48	77	39	88

Exercise 3

1. 192	175	180	531	1708
2. 1548	2754	3269	4450	2900

Exercise 4

1. 240	420	1150	2100	1820
2. 8508	13,984	14,245	22,088	7524

Exercise 7

1. $144 \div 12$	**2.** 30×100	**3.** $120 - 60$	**4.** $30 + 40$
5. $25 - 5$	**6.** 100×5	**7.** $200 \div 20$	**8.** $70 \div 5$

Exercise 8

1. 32	27	21	22	44
2. 24	275	317	872	214

Exercise 9

1. 58 R1	25 R2	31 R3	24 R2	22 R3
2. 28 R4	21 R1	23 R1	25 R6	46 R8

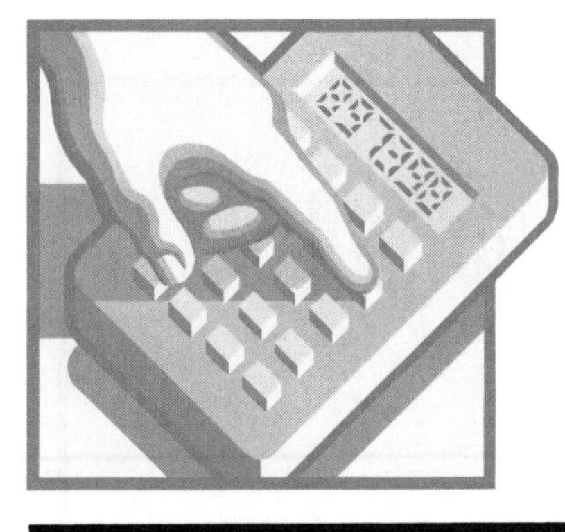

LESSON 4

Finding Amounts by Multiplying and Dividing Decimals

Imagine going to the store to buy six pens at $.99 each. Would $.99 be added six times to find the total? Perhaps, but it would be much faster to multiply. If buying a roll of tape and the sign in the store says, "3 rolls for $.99," how would the price of one roll be calculated? By division.

Finding amounts by multiplying and dividing decimals is very common in business. Multiplication and division of amounts of money involve calculating decimal amounts. Finding earnings for the week, placing an order with another company, and calculating a customer's bill all involve decimal amounts.

Katy Kelly solves the following work problem.

Work Problem

When City Office Supply has a sale, they send a flier and an order form to their regular customers. Katy Kelly is a secretary at the Thompson and Thompson Law Offices. She received City Office Supply's most recent mailing, which is shown on the next page. Katy wanted to order 30 legal pads and 6 boxes of computer diskettes for the office. She paid careful attention to how the sales prices were listed on the flier so she could complete the order form correctly.

City Office Supply

is having a sale!

Office Supplies

ITEM#	ITEM	SALE PRICE
BP63	Black Pens	12 for $1.44
CT19	Cellophane Tape	2 rolls for $1
CE34	Clasp Envelopes	2 boxes for $12
CD30	Computer Diskettes (box of 10)	$14.95 per box
CP24	Computer Paper	$25.50 per case
PC57	Copy Paper	$4.95 per ream
SE84	Envelopes (box of 50)	2 boxes for $1.50
TR06	Expense and Tax Record	$5.50 each
FP50	Felt Tip Pens	2 for $1
FF62	File Folders (box of 50)	2 boxes for $15
LP71	Legal Pads	12 for $6
MP15	Memo Pads	12 for $1.80
WP37	Pencils, #2	12 for $ .96
PS08	Scissors	$1.99 each
SS29	Stapler	$5.95 each

Furniture

EC75	Executive Chair	$339.95
FC44	Four-drawer Filing Cabinet	$119.95
SC52	Storage Cabinet	$139.95
CC80	Steno Chair	$109.95
FC24	Two-drawer Filing Cabinet	$79.95

Free delivery on any metro order over $100!

ORDER FORM

City Office Supply
1000 Clarke Road
Memphis, TN 38115

Ship to: Thompson & Thompson
2929 James Rd.
Memphis, TN 38117

ITEM #	ITEM	QTY.	UNIT	UNIT PRICE	TOTAL PRICE
LP71	Legal Pads	30	each		
CD30	Computer Disks	6	boxes	14.95	

TOTAL AMOUNT DUE:

The sale price for each item on the flier provided information about the unit for ordering the item. For example, computer diskettes were $14.95 a box, so Katy knew to order by the box. She knew that legal pads could be ordered individually because there was no label, such as "box," in the sale price.

Katy calculated the *unit price* (the price of one unit) of the legal pads by dividing 12 into $6. Divide below and write the price of one legal pad on the form above.

$$12\overline{)6.00}$$

The unit price of legal pads is $.50. To find the total price for 30 pads, Katy multiplied the quantity by the unit price ($.50). Multiply 30 × $.50 and write the answer on the form.

$$\begin{array}{r} \$.50 \\ \times 30 \\ \hline \end{array}$$

Katy found that the total price for 30 legal pads was $15. Then she calculated the total price of the six boxes of computer diskettes. She multiplied the quantity (6) by the unit price ($14.95). Multiply, and write the total price of the computer diskettes on the form.

$$\begin{array}{r} \$14.95 \\ \times \quad 6 \\ \hline \end{array}$$

Katy found the total price of the computer diskettes to be $89.70. Finally, Katy added the two amounts to find the *total amount due*. Add and write the answer on the form on the previous page.

$15.00
+89.70

The completed form should appear as shown below. The total amount due was $104.70. Students whose answers did not agree with Katy's might need more practice in multiplying and dividing decimals. The section on page 107 provides assistance and practice multiplying and dividing decimals.

ORDER FORM

City Office Supply
1000 Clarke Road
Memphis, TN 38115

Ship to: Thompson & Thompson
2929 James Rd.
Memphis, TN 38117

ITEM #	ITEM	QTY.	UNIT	UNIT PRICE	TOTAL PRICE
LP71	Legal Pads	30	each	.50	15.00
CD30	Computer Disks	6	boxes	14.95	89.70
				TOTAL AMOUNT DUE:	104.70

Doing Math to Find Amounts

Katy knew she was looking for an amount, the total amount due. Katy planned her solution:

1. She needed the total price for each item.

2. Using the information on the sales flier, she could calculate the total prices.

3. She knew how many of each item to order. She also knew the unit price for the diskettes.

4. She planned to divide to find the unit price of the legal pads, multiply to find the total prices of the legal pads and computer diskettes, and add to find the total amount due.

Katy solved the problem by following the calculation steps three times. The first time she found the unit price for the legal pads. The second time she found the total price of the items she wanted to order. The third time she found the total amount due.

She first calculated to find the unit price of the legal pads.

1. Select the relevant data. *12 pads for $6.00*

2. Set up the calculation.

$$12 \overline{)6.00}$$

3. Do the calculation.

$$\begin{array}{r} .50 \\ 12 \overline{)6.00} \\ \underline{60} \\ 0 \end{array}$$

4. Check the accuracy of the answer.

$$\begin{array}{r} \$\ .50 \\ \underline{\times\ 12} \\ \$6.00 \end{array}$$

Each legal pad cost $.50.

Next, Katy calculated the total prices.

1. Select the relevant data. *Quantities and unit prices for the legal pads and computer diskettes*

2. Set up the calculations.

Legal Pads	*Diskettes*
$ .50	$14.95
× 30	× 6

3. Do the calculations.

Legal Pads	*Diskettes*
$15.00	$89.70

4. Check the accuracy of the answers.

$$30\overline{)15.00} = .50 \quad (15\,0,\ 0)$$

$$6\overline{)89.70} = 14.95 \quad (6,\ 29,\ 24,\ 57,\ 54,\ 30,\ 30,\ 0)$$

The legal pads cost $15.00, and the computer diskettes cost $89.70.

Finally, Katy calculated the total amount due.

1. Select the relevant data. *Total costs for the legal pads and the computer diskettes*

2. Set up the calculation.
   $ 15.00
   + 89.70

3. Do the calculation. $104.70

4. Check the accuracy of the answer.
   $104.70
   − 89.70
   $ 15.00

The total amount due for the order was $104.70.

Now Marie Palumbo, another customer of City Office Supply, works through an entire problem.

Work Problem

Marie is the secretary at Oak Park School. One morning the principal came to Marie and said, "We need to order two new four-drawer filing cabinets for the office. Would you please take care of that?"

Marie is also in charge of the central supplies at Oak Park, and she noticed that the pencil supply was low. When Marie received the sales flier from City Office Supply, she decided to order both the file cabinets and the pencils. She wrote the information she knew on this order form. Then she solved the problem to find the missing information. Notice that she worked through the SOLVE steps three times.

ORDER FORM

City Office Supply
1000 Clarke Road
Memphis, TN 38115

Ship to: Oak Park School
300 Elm Street
Memphis, TN 38117

ITEM #	ITEM	QTY.	UNIT	UNIT PRICE	TOTAL PRICE
WP37	#2 Pencils	200	each		
FC44	4 Drwr Filing Cabinet	2	each		

TOTAL AMOUNT DUE:

DEFINE the problem

- What is the expected outcome? *The total amount due*
- What is the purpose? *To find an amount*

PLAN the solution

- What data are needed? *The total prices for the filing cabinet and pencils*

- **Where can the data be found?** *From calculations with amounts on the sales flier and the order form*

- **What is already known?** *Quantities to be ordered and the unit price of the filing cabinet*

- **Which operations should be used?** *Division, multiplication and addition*

SOLVE the problem

The first calculation was to find the unit price of the pencils.

- Select the relevant data. *12 pencils cost $1.44*

- Set up the calculation.

$$12\overline{)1.44}$$

- Do the calculation.

$$\begin{array}{r} .12 \\ 12\overline{)1.44} \\ \underline{12} \\ 24 \\ \underline{24} \\ 0 \end{array}$$

- Check the accuracy of the answer.

$$\begin{array}{r} 12 \\ \underline{\times .12} \\ \$1.44 \end{array}$$

Next, Marie calculates the total prices of the pencils and the filing cabinets.

- Select the relevant data. *Quantities and unit prices for pencils and filing cabinets*

- Set up the calculation.

$$\begin{array}{r} 200 \\ \times\ .12 \\ \hline \end{array} \qquad \begin{array}{r} 119.95 \\ \times\quad 2 \\ \hline \end{array}$$

- Do the calculation. $24.00 $239.90

- Check the accuracy of the answer.

$$.12\overline{)24} = 200;\quad 24,\ 0$$

$$2\overline{)239.90} = 119.95;\quad 2,\ 3,\ 2,\ 19,\ 18,\ 19,\ 18,\ 10,\ 10,\ 0$$

Finally, Marie calculated the total amount due.

- Select the relevant data. *Total amounts for the pencils and filing cabinets*

- Set up the calculation.

$$\begin{array}{r} \$239.90 \\ +\ 24.00 \\ \hline \end{array}$$

- Do the calculation. $263.90

- Check the accuracy of the answers.

$$\begin{array}{r} \$263.90 \\ -\ 24.00 \\ \hline \$239.90 \end{array}$$

CHECK the solution

Make sure the calculation solves the work problem.

- Was the defined purpose accomplished? *Yes, the total amount due is $263.90*

- Is the solution to the work problem reasonable? *Yes, $263.90 is reasonable because it is greater than both of the total prices*

The completed form should appear as shown below:

ORDER FORM

City Office Supply
1000 Clarke Road
Memphis, TN 38115

Ship to: Oak Park School
300 Elm Street
Memphis, TN 38117

ITEM #	ITEM	QTY.	UNIT	UNIT PRICE	TOTAL PRICE
WP37	#2 Pencils	200	each	.12	24.00
FC44	4 Drwr Filing Cabinet	2	each	119.95	239.90

TOTAL AMOUNT DUE: $263.90

Problem-Solving Practice

Use the DEFINE, PLAN, SOLVE, and CHECK steps to solve the next work problem. Write the answers on the order form. The item number, item description, and unit are already on the form.

Work Problem

Ralph is the office assistant at Tots and Toddlers Day Care Center. The owner and head teacher, Claire Jackson, says, "I see from this flier that City Office Supply is having a sale. We need more cellophane tape. We also need a storage cabinet for those new puzzles and blocks. Please order fifteen rolls of tape and a new storage cabinet. Here's what you need." Then she gave Ralph an order form and the flier shown at the beginning of this lesson. Complete the order form for Ralph.

ORDER FORM

City Office Supply
1000 Clarke Road
Memphis, TN 38115

Ship to: Tots & Toddlers Day
132 Sandbox Lane
Memphis, TN 38121

ITEM #	ITEM	QTY.	UNIT	UNIT PRICE	TOTAL PRICE
CT19	Cellophane Tape		rolls		
SC52	Storage Cabinet		each		

TOTAL AMOUNT DUE:

DEFINE the problem

- What is the expected outcome? ______________________________

- What is the purpose? ______

PLAN the solution

- What data are needed? ______
- Where can the data be found? ______
- What is already known? ______
- Which operations should be used? ______

SOLVE the problem

- Select the relevant data. ______
- Set up the calculation.
- Do the calculation.

- Check the accuracy of the answer.

Work through the SOLVE steps again if necessary.

CHECK the solution

Make sure the calculation solves the work problem.

- Was the defined purpose accomplished? ____________________

- Is the solution to the work problem reasonable? ____________________

Answers to Problem-Solving Practice questions appear on page 115.

On Your Own

Solve the following work problems. Remember to DEFINE the problem, PLAN the solution, SOLVE the problem, and CHECK the solution to make sure it solves the work problem. Write the answers on the order form. The item number, item description, and unit are already included on the form.

Work Problem A

The office clerk at Omen Graphics is a regular customer of City Office Supply. Part of her job is to monitor general supplies and order more when needed. When the flier and order form from City Office Supply come in the morning mail, she decides to take advantage of the sale and order eight cases of computer paper and 25 felt tip pens.

ORDER FORM

City Office Supply
1000 Clarke Road
Memphis, TN 38115

Ship to: Omen Graphics
107 E. 53rd Street
Memphis, TN 38117

ITEM #	ITEM	QTY.	UNIT	UNIT PRICE	TOTAL PRICE
CP24	Computer Paper		case		
FP50	Felt Tip Pens		each		

TOTAL AMOUNT DUE:

Work Problem B

Gloria Sanchez is the secretary at Eastside Community Services. The director, Jim Serano, brings her the sales flier and order form and says, "This sales flier from City Office Supply reminds me that we need a few things. Would you take care of the ordering? We could use 30 black pens and some scissors. Order three pairs. We could use some copier paper, too. Order enough paper for ten weeks." Gloria knows from experience that the office uses two reams of paper per week.

ORDER FORM

City Office Supply
1000 Clarke Road
Memphis, TN 38115

Ship to: Eastside Community Services
579 E. 33rd St.
Memphis, TN 38117

ITEM #	ITEM	QTY.	UNIT	UNIT PRICE	TOTAL PRICE
BP63	Black Pens		each		
PC57	Copy Paper		ream		
PS08	Scissors		each		

TOTAL AMOUNT DUE:

Work Problem C

The owner of Beedle Exterminators, Jerry Beedle, comes to the office clerk and says, "I was in the supply area today, and I noticed we are running short of a few things. Please order what we need. Three more boxes of clasp envelopes should last us through the year. We need memo pads, too. We use about five pads a week. Get enough to last eight weeks. I also need two new filing cabinets for my office. Order the two-drawer model." Solve this problem and complete the order form.

ORDER FORM

City Office Supply
1000 Clarke Road
Memphis, TN 38115

Ship to:
Beedle Exterminators
33 Termite Terrace
Memphis, TN 38116

ITEM #	ITEM	QTY.	UNIT	UNIT PRICE	TOTAL PRICE
CE34	Clasp Envelopes		box		
MP15	Memo Pads		each		
FC24	2-Drawer Filing Cabinet		each		

TOTAL AMOUNT DUE:

Work Problem D

The purchasing manager at Write On Stationery, often gets merchandise from City Office Supply. He decides to take advantage of the sale. Order these items: 75 boxes of envelopes, 10 dozen black pens, 36 pairs of scissors, and 25 staplers.

ORDER FORM

City Office Supply
1000 Clarke Road
Memphis, TN 38115

Ship to: Write On Stationery
444 Covington Place
Memphis, TN 38122

ITEM #	ITEM	QTY.	UNIT	UNIT PRICE	TOTAL PRICE
SE84	Envelopes		box		
BP63	Black Pens		each		
PS08	Scissors		each		
SS29	Staplers		each		

TOTAL AMOUNT DUE:

Work Problem E

Joyce Allen is the president of Ride & Roll Sports Store. Ride & Roll had a good year last year, so they will be expanding. Joyce asks, "Please order the furniture for our new offices. We need two more executive chairs, 4 four-drawer filing cabinets, 3 stenographer chairs, and 4 two-drawer filing cabinets. Order from City Office Supply."

Complete the order form.

ORDER FORM

City Office Supply
1000 Clarke Road
Memphis, TN 38115

Ship to:
Ride & Roll Sports Store
313 Loan Oak Road
Memphis, TN 38117

ITEM #	ITEM	QTY.	UNIT	UNIT PRICE	TOTAL PRICE
EC75	Executive Chair		each		
FC44	4-Drawer Filing Cabinet		each		
CC80	Steno Chair		each		
FC24	2-Drawer Filing Cabinet		each		
			TOTAL AMOUNT DUE:		

Skills Practice: Multiplying and Dividing Decimals

MULTIPLYING DECIMALS

Multiplying decimals begins the same as multiplying whole numbers. Line up the digits on the right. (Notice that, unlike adding and subtracting decimals, the decimal points do not have to be lined up.) After multiplying, count the number of decimal places in the problem and put the same number of decimal places in the answer.

Step 1. Line up the digits on the right.

```
  3.15
× 1.2
```

Step 2. Multiply as usual.

```
   1
  3.15
× 1.2
   630
  315
  3780
```

Step 3. Count the number of decimal places in the problem. Put the same number of decimal places in the answer.

```
   1
  3.15   2 places
× 1.2    1 place
   630
  315
 3.780   3 places
```

Count three places from the right. Put the decimal point between the 3 and the 7.

Sometimes a zero must be added to the answer as shown in the problem below. The answer needs five, but there are only four digits.

```
   .178   3 places
×   .15   2 places
    890
   178
 .02670   5 places
```

Add a zero to the left of the answer. The zero is referred to as a placeholder.

Exercise 1

Multiply.

1. $\begin{array}{r}\$3.60\\ \times\ 15\\ \hline\end{array}$ $\begin{array}{r}\$23.56\\ \times\ 35\\ \hline\end{array}$ $\begin{array}{r}\$35.21\\ \times\ 23\\ \hline\end{array}$ $\begin{array}{r}\$45.89\\ \times\ 48\\ \hline\end{array}$

2. $\begin{array}{r}47.84\\ \times\ 3.4\\ \hline\end{array}$ $\begin{array}{r}2.436\\ \times\ 25\\ \hline\end{array}$ $\begin{array}{r}0.1783\\ \times\ 1.6\\ \hline\end{array}$ $\begin{array}{r}0.4726\\ \times\ 27\\ \hline\end{array}$

Answers begin on page 116.

Exercise 2

Multiply.

1. $\begin{array}{r}478\\ \times 0.47\\ \hline\end{array}$ $\begin{array}{r}63.7\\ \times 0.25\\ \hline\end{array}$ $\begin{array}{r}592\\ \times 0.0015\\ \hline\end{array}$ $\begin{array}{r}42.8\\ \times 0.07\\ \hline\end{array}$

2. $\begin{array}{r}24.5\\ \times 0.43\\ \hline\end{array}$ $\begin{array}{r}31.6\\ \times 0.57\\ \hline\end{array}$ $\begin{array}{r}3.000\\ \times 0.035\\ \hline\end{array}$ $\begin{array}{r}2.345\\ \times 0.062\\ \hline\end{array}$

DIVIDING DECIMALS

The same four division steps used in Lesson 3 are used when the division process includes a decimal. Review these examples.

Example 1

Dividing a whole number by another whole number to get a decimal:

$$20\overline{)\,8}$$

Step 1. Before dividing, put the decimal point in the dividend. Put another decimal point directly above the first, on the line where the quotient will go. Add a zero to the right of the decimal point in the dividend.

$$\begin{array}{r} . \\ 20\overline{)\,8.0} \end{array}$$

Put the decimal points here.
Add a zero to the right of the decimal point.

Step 2. Divide as usual.

$$\begin{array}{r} .4 \\ 20\overline{)\,8.0} \\ \underline{8\,0} \\ 0 \end{array}$$

Example 2

Dividing a decimal by a whole number:

Follow the same steps as shown in Example 1. Before starting, put a decimal point on the line where the answer will go. Be sure the new decimal point is directly above the decimal point in the dividend. Add zeros, as needed, to the right of the last decimal place of the dividend.

$$\begin{array}{r} .05 \\ 150\overline{)7.50} \\ \underline{7\,5} \\ 0 \end{array}$$

Put a decimal point here.
Add a zero here.
Divide as usual.

Example 3

Dividing a decimal by another decimal:

When the divisor contains a decimal, move the decimal point to the right to turn the divisor into a whole number. Count the number of places the decimal point is moved. Then move the decimal point that same number of places in the dividend. Place the decimal point for the quotient directly above the new location of the decimal point in the dividend.

$$1.2\overline{)14.4}$$

Step 1. Move the decimal one place to the right in the divisor to make it a whole number. Then move the decimal point the same number of places (in this case, one place) to the right in the dividend. Put the decimal point for the quotient on the top line.

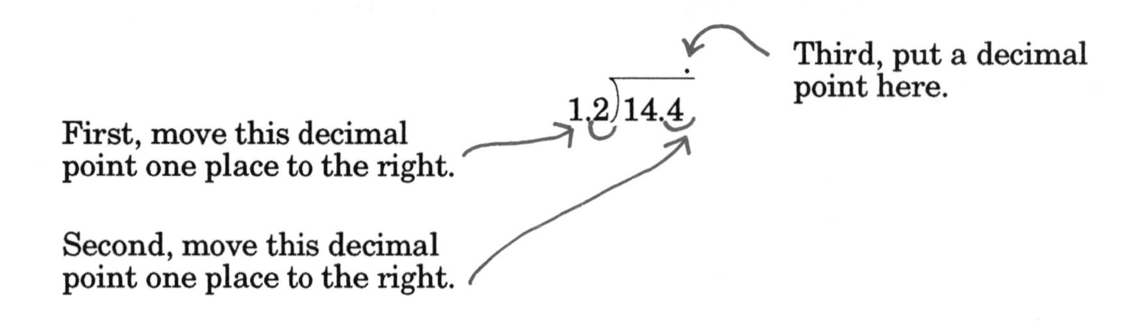

Step 2. Divide as usual.

```
      12.
12)  144.
     12
      24
      24
       0
```

Example 4

Dividing a whole number by a decimal:

Step 1. As in Example 3, set up the problem first. Move the decimal point in the divisor to make a whole number. Move the decimal point in the dividend the same number of places to the right. Add zeros to the right of the dividend. Put a decimal point above the dividend in the answer.

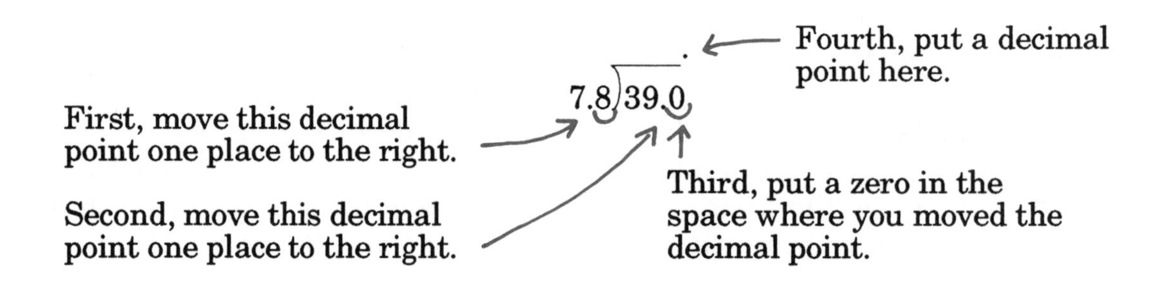

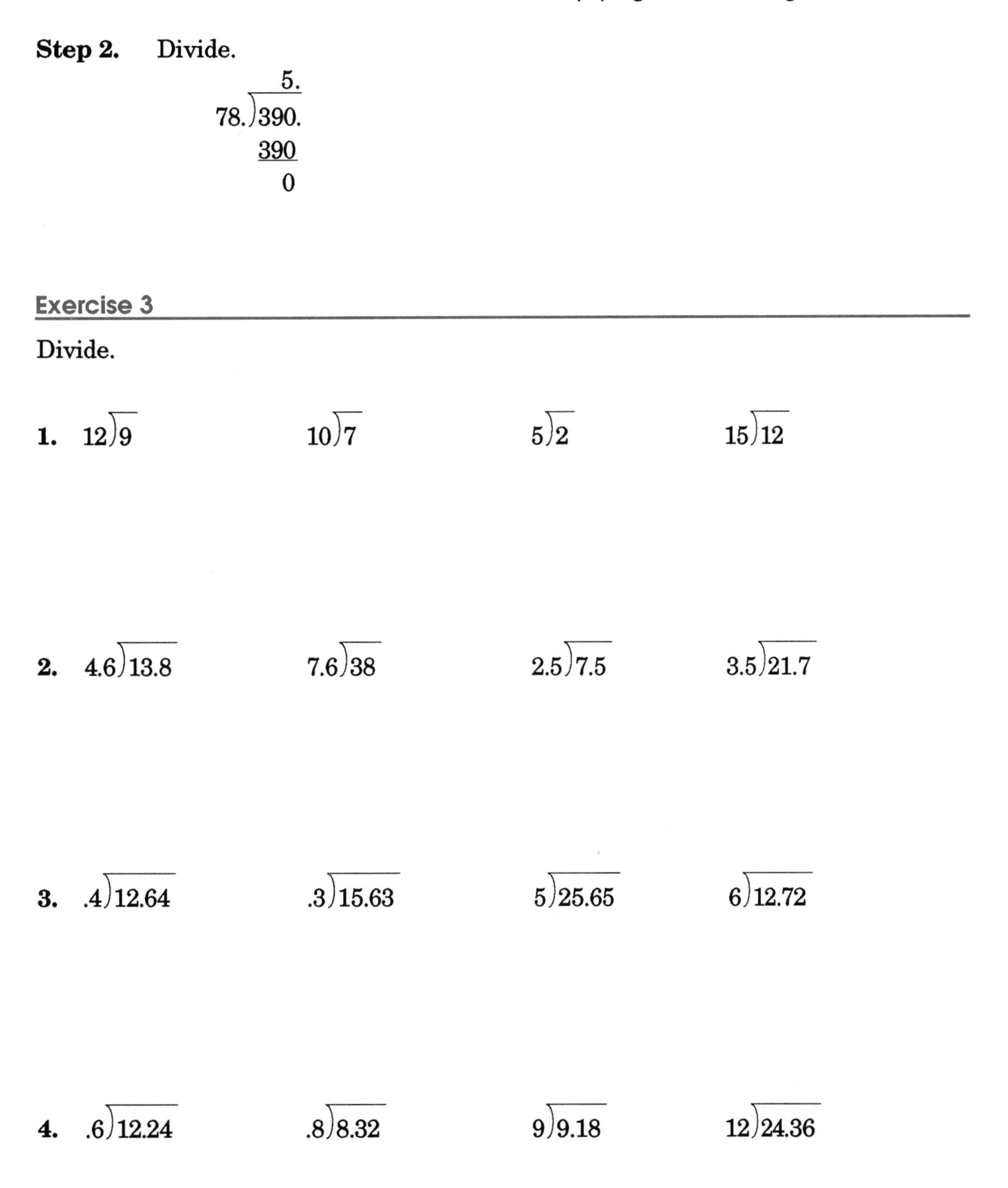

Step 2. Divide.

$$\begin{array}{r} 5. \\ 78.\overline{)390.} \\ \underline{390} \\ 0 \end{array}$$

Exercise 3

Divide.

1. $12\overline{)9}$ $\quad 10\overline{)7}$ $\quad 5\overline{)2}$ $\quad 15\overline{)12}$

2. $4.6\overline{)13.8}$ $\quad 7.6\overline{)38}$ $\quad 2.5\overline{)7.5}$ $\quad 3.5\overline{)21.7}$

3. $.4\overline{)12.64}$ $\quad .3\overline{)15.63}$ $\quad 5\overline{)25.65}$ $\quad 6\overline{)12.72}$

4. $.6\overline{)12.24}$ $\quad .8\overline{)8.32}$ $\quad 9\overline{)9.18}$ $\quad 12\overline{)24.36}$

Answers begin on page 116.

Exercise 4

Divide.

1. $8\overline{)1.2}$ $5\overline{)2.1}$ $6\overline{)3.3}$ $9\overline{)3.6}$ $4\overline{)1.1}$

2. $.3\overline{)1.8}$ $.6\overline{)2.22}$ $.8\overline{)3.02}$ $.4\overline{)1.162}$

3. $.8\overline{)27}$ $.6\overline{)27}$ $.4\overline{)1}$ $.8\overline{)98}$

4. $40\overline{)1.04}$ $20\overline{)0.7}$ $8\overline{)0.2}$ $4\overline{)0.01}$

Check Yourself

Circle the letter of each correct answer.

1. To find the total price of an item, ____________________

a. add the quantity and the unit price.

b. divide the quantity into the unit price.

c. divide the unit price into the quantity.

d. multiply the quantity by the unit price.

2. To find the total amount due on an order form, ____________________

a. add the total prices.

b. multiply the quantity by the total price.

c. multiply the quantity by the unit price.

d. divide the total price by the quantity.

3. Which of these situations requires division?

a. Calculating total pay for the week

b. Buying one package of index cards priced at 3 for $2.00

c. Finding the total price of 5 identical notebooks

d. Both b and c

4. Which of these situations requires multiplication?

a. Finding the average number of miles to the gallon

b. Finding the unit price of an item priced at 2 for $.25

c. Calculating the total price for 8 boxes of markers

d. Both b and c

5. City Office Supply sells markers for $.99 each. Katy needed 20 pens, so she paid $19.80. City Office Supply packaged the pens in boxes of 10 and mailed them to her. What is the unit price of the pens?

a. $.99

b. $20.00

c. $19.80

d. $10.00

Work Problem

Norman Glass, of Glass Windows, received the sales flier from City Office Supply. He says, "Order a few things from this list. Buy two expense and tax records, 2 two-drawer filing cabinets, and enough file folders to last ten weeks. We usually use one box every two weeks." Use the information on the sales flier at the beginning of this lesson to solve this work problem and complete the order form for Norman.

ORDER FORM

City Office Supply
1000 Clarke Road
Memphis, TN 38115

Ship to: Glass Windows
3790 Payne Ave.
Memphis, TN 38116

ITEM #	ITEM	QTY.	UNIT	UNIT PRICE	TOTAL PRICE
TR06	Expense & Tax Record		each		
FF62	File Folders		box		
FC24	Two-Drawer Filing Cabinet		each		

TOTAL AMOUNT DUE:

Answers to Problem-Solving Practice Questions

DEFINE the problem

- the total amount due
- to find an amount

PLAN the solution

- the total prices for the tape and the storage cabinet
- from the calculations with the amounts on the flier and the order form
- quantities to be ordered and the unit price of the filing cabinet
- division, multiplication and addition

SOLVE the problem

Calculate the unit price of the tape.

- Two rolls of tape sell for $1.00.
- 1 ÷ 2 =
- $.50
- 2 × $.50 = $1

Calculate the total price of the tape.

- Fifteen rolls of tape are needed; tape costs $.50 a roll.
- 15 × .50 =
- $7.50
- 7.50 ÷ .50 = 15

Calculate the total amount due.

- The tape costs $7.50. The storage cabinet costs $139.95.
- $ 7.50
+139.95
- $ 147.45
- $ 147.45
−139.95
$ 7.50

CHECK the solution

- The defined purpose was accomplished; the total amount due is $147.45.
- The solution to the work problem, $147.45, is reasonable because it is greater than each of the total prices.

ORDER FORM

City Office Supply
1000 Clarke Road
Memphis, TN 38115

Ship to: Tots & Toddlers Day
132 Sandbox Lane
Memphis, TN 38117

ITEM #	ITEM	QTY.	UNIT	UNIT PRICE	TOTAL PRICE
CT19	Cellophane Tape	15	rolls	.5	7.50
SC52	Storage Cabinet	1	each	139.95	139.95
				TOTAL AMOUNT DUE:	$147.45

Answers to Skills Practice Problems

Exercise 1

1.	$54	$824.60	$809.83	$2,202.72
2.	162.656	60.9	.28528	12.7602

Exercise 3

1.	.75	.7	.4	.8
2.	3	5	3	6.2
3.	31.6	52.1	5.13	2.12
4.	20.4	10.4	1.02	2.03

Putting It All Together

In these two work problems, students will combine the different skills and knowledge presented in this unit. Students learned the four problem-solving strategy steps: DEFINE, PLAN, SOLVE, and CHECK. Various office forms were used in problem solving activities. Students learned to use the information on forms and how to complete them. Completing a form often means finding amounts. Students learned to find amounts by adding, subtracting, multiplying, and dividing whole numbers and decimals.

Solve the following work problems. Remember to look and listen for words such as "total" and phrases such as "how many in each" to help decide which operation to use.

Work Problem A

Adam Royce works in the accounting department at City Office Supply. Adam worked 40 regular hours plus 5 overtime hours per week during the pay period. He earns $9.34 an hour for both regular and overtime hours. A pay period has two weeks. Complete the payroll slip below. Be sure to show Adam's net pay.

CITY OFFICE SUPPLY

PAYROLL STATEMENT

Employee *Adam Royce*

Employee No. *307* Department *560*

Payroll Classification *B-2* S. Sec. No.

PAYROLL PERIOD From *February 1, 199–* To *February 14, 199–*

EARNINGS			DEDUCTIONS		
Salary			F.I.C.A.	54	64
Overtime			Fed'l W.H. Tax	97	47
Tips			Unemployment Ins.	42	03
			State W.H. Tax	33	62
Total Earnings			Total Deductions		
Net Pay					

Work Problem B

A new semester begins soon, and many teachers are sending requisitions for supplies. The Oak Park School is out of pencils, memo pads, computer disks and cellophane tape. Place an order with City Office Supply. Use the order form on the next page to order enough supplies to fill the three teachers' requisitions shown below. Refer to the sale flier on page 90.

OAK PARK SCHOOL SUPPLIES REQUISITION

ROOM # 132 NAME Carlos Gomez DATE 12/9

ITEM #	QUANTITY	UNIT	DESCRIPTION
CT19	4	rolls	Cellophane Tape
MP15	6	each	Memo Pads
WP37	25	each	#2 Pencils

OAK PARK SCHOOL SUPPLIES REQUISITION

ROOM # 213 NAME Chris Bailey DATE 12/2

ITEM #	QUANTITY	UNIT	DESCRIPTION
CD30	3	box	Computer Diskettes
MP15	8	each	Memo Pads
WP37	30	each	#2 Pencils

OAK PARK SCHOOL SUPPLIES REQUISITION

ROOM # 321 NAME Ellen Kowalski DATE 12/9

ITEM #	QUANTITY	UNIT	DESCRIPTION
CT19	5	rolls	Cellophane Tape
CD30	2	box	Computer Diskettes
WP37	24	each	#2 Pencils

ORDER FORM

City Office Supply
1000 Clarke Road
Memphis, TN 38115

Ship to: Oak Park School
300 Elm Street
Memphis, TN 38117

ITEM #	ITEM	QTY.	UNIT	UNIT PRICE	TOTAL PRICE
CT19	Cellophane Tape		roll		
CD30	Computer Disks		box		
MP15	Memo Pads		each		
WP37	#2 Pencils		each		

TOTAL AMOUNT DUE:

UNIT II

EXPRESSING RELATIONSHIPS

Office workers often use *fractions*, *percents*, or *ratios* on the job. Fractions, percents, and ratios express relationships between two numbers. *Fractions* relate a part or parts to the whole. *Percents* are similar to fractions, but they relate a select number of parts to one hundred. *Ratios* compare two different kinds of amounts. Expression of a ratio usually includes words such as *to*, *per* or *for every*.

To express a number relationship as a fraction, ratio, or percent, these steps must be followed:

1. Select the data.
2. Set up the calculation.
3. Do the calculation.
4. Check the accuracy of the answer.
5. Express the relationship in desired terms.

The first three lessons in this unit explain how to express relationships between amounts found on income statements, enrollment records and expense summaries. The fourth lesson explains how to use percents to calculate pay raises.

Neighborhood Youth and Family Services

The setting for activities in this unit is the office at Neighborhood Youth and Family Services. This is a community organization that offers day care for children, recreational activities for high school youth, and classes for adults. The director is Ted Urbanski. Cindy Spears is the receptionist and office assistant. Darren Simms takes care of the bookkeeping and payroll records. Other employees teach special classes and coordinate the various programs.

LESSON 5

Expressing Relationships as Fractions

Fractions indicate how one or more parts relate to the whole. Office workers must express relationships as fractions when they are asked to find *what part*. They may need to know what part of the inventory has been sold or what part of the expenses are used for a particular purpose. They may also need to determine what part of the *income*, or money that is received, comes from a particular source, as shown in the problem below.

Cindy Spears' solution to the following work problem is shown as an example.

Work Problem

Ted was working on his annual report to the board of directors. He asked Cindy to find out what part of Neighborhood Youth and Family Services' income came from state grants. Cindy used the data in the income statement below to answer Ted's question. (An *income statement* lists the sources of income and tells how much money the business or organization has received from each source.)

Yearly Income Statement
Neighborhood Youth and Family Services

Child Care:	
ages Birth - 5	$120,000
ages 6 - 12	$60,000
Rental Fees (Pool and Gym):	$20,000
Class and Recreation Fees:	$10,000
Grants:	
private	$40,000
state	$80,000
federal	$30,000
TOTAL INCOME:	$360,000

The part of the income that came from state grants was $80,000. The whole, or total income, was $360,000. Cindy set up the fraction as shown below and planned to reduce it to lowest terms. Write the reduced fraction after the equal sign.

$$\frac{80{,}000}{360{,}000} =$$

Cindy found that 2/9 of the year's income came from state grants. Students who found a fraction other than 2/9 may need more practice in reducing fractions. The section that begins on page 132 provides assistance with fractions.

Doing Math to Express Relationships

Cindy defined her purpose as expressing a relationship. From her plan, she knew that she needed the amount of the agency's total income and the amount that it received from state grants. She found these amounts on the income statement. She planned to reduce the fraction to lowest terms.

Then Cindy followed five steps to solve the problem.

1. Select the relevant data.

   Fractions relate the part to the whole. The part was the amount that the agency received from state grants, $80,000. The whole was the agency's total income, $360,000.

2. Set up the calculation.

   Fractions are set up with the part above the whole, so Cindy set up her calculation as shown below:

$$\frac{80{,}000}{360{,}000}$$

3. Do the calculation.

Cindy reduced the fraction to lowest terms in two steps. First she divided the numerator and denominator by 10,000 to eliminate the zeros. Then she divided the numerator and denominator by 4.

$$\frac{80,000}{360,000} = \frac{8}{36} = \frac{2}{9}$$

4. Check the accuracy of the answer.

Cindy cross multiplied.

$$\frac{9}{2} \times \frac{80,000}{360,000} = \frac{720,000}{720,000}$$

The cross products are equal, so Cindy calculated correctly.

5. Express the relationship in desired terms.

Three ways that number relationships can be expressed are fractions, ratios, and percents. Ted asked her to find a part, so she wrote the answer as a fraction.

$$\frac{2}{9}$$

The state grants provided 2/9 of the year's income.

Now Cindy works through an entire problem.

Work Problem

Ted needed more information for his report. He asked Cindy, "What part of our total income comes from the child care program for children aged five and under?"

DEFINE the problem

- What is the expected outcome? *A part of the total income*
- What is the purpose? *To express a relationship*

Note: When the expected outcome is a part of some whole, the purpose is to express a relationship.

PLAN the solution

- What data are needed? *Income from the child care program for children under five and the total income*
- Where can the data be found? *On income statement*
- What is already known? *Income from the child care program for children under five and the total income*
- Which operation should be used? *Division*

SOLVE the problem

- Select the relevant data.

  Income from the child care program for children under five and the total income.

- Set up the calculation.

  $$\frac{120{,}000}{360{,}000}$$

- Do the calculation.

  $$\frac{120{,}000}{360{,}000} = \frac{12}{36} = \frac{1}{3}$$

- Check the accuracy of the answer.

  $$\frac{3}{1} \times \frac{120{,}000}{360{,}000} = \frac{360{,}000}{360{,}000}$$

- Express the relationship in desired terms.

  $$\frac{1}{3}$$

CHECK the solution

Make sure the calculation solves the work problem.

- Was the defined purpose accomplished?

  Yes, 1/3 expresses a relationship

- Is the solution to the work problem reasonable?

  Yes, 1/3 is reasonable because it is less than 1

  Note: Cindy's expected outcome was a part of a whole. A part of a whole is a fraction less than one. Cindy expected her answer to be less than one.

The child care program for children five and under provided 1/3 of the year's income.

Problem-Solving Practice

Use the DEFINE, PLAN, SOLVE, and CHECK steps to solve the next problem.

Work Problem

Cindy is helping Ted prepare a graph to show the sources of income for Neighborhood Youth and Family Services. Ted says, "What part of the income comes from the rental of the pool and gym?" Use the income statement in the beginning of this lesson to answer Ted's question.

DEFINE the problem

- What is the expected outcome? ______________________________

____

____

- What is the purpose? ______________________________

____

____

PLAN the solution

- What data are needed? ______________________________

____

____

- Where can the data be found? ______________________________

____

____

- What is already known? ______________________________

____

____

- Which operation should be used? ______________________________

SOLVE the problem

- Select the relevant data. ____________________

- Set up the calculation.

- Do the calculation.

- Check the accuracy of the answer.

- Express the relationship in desired terms. ____________________

CHECK the solution

Make sure the calculation solves the work problem.

- Was the defined purpose accomplished? ____________________

- Is the solution to the work problem reasonable? ____________________

- What part of the income comes from the rental of the pool and gym? ____________________

Answers to Problem-Solving Practice questions appear on page 137.

On Your Own

Here are some work problems for added practice. Remember to DEFINE the problem, PLAN the solution, SOLVE the problem, and CHECK the solution. Make sure the solution solves the work problem. In all of the problems, Ted is preparing information for the yearly report to the board.

Use the income statement at the beginning of the lesson for Work Problems A and B.

Work Problem A

Ted would like to know what part of the income comes from class and recreation fees.

Answer: ____________________

Work Problem B

Ted wants to know what part of the income comes from all the child care services that are provided.

Answer: ____________________

Work Problem C

One year later, Ted is preparing another annual report. He asks, "What part of our income is provided by the federal grant this year?" Use the income statement shown below.

Answer: ____________________

Use the following income statement for Work Problems D and E.

Yearly Income Statement
Neighborhood Youth and Family Services

Child Care:	
ages Birth - 5	$120,000
ages 6 - 12	$50,000
Rental Fees (Pool and Gym):	$20,000
Class and Recreation Fees:	$10,000
Grants:	
private	$80,000
state	$80,000
federal	$40,000
TOTAL INCOME:	$400,000

Work Problem D

Ted says, "Please find out what part of our income came from private grants."

Answer: ______________________________

Work Problem E

Ted asks, "What part of the total income is supplied by all the grants?"

Answer: ______________________________

Skills Practice: Fractions

UNDERSTANDING FRACTIONS

One way to express a relationship between numbers is to compare the part to the whole. Such a comparison is called a *fraction*. The circle below is divided into four equal parts. The one shaded part is compared to the 4 equal parts. In other words, 1/4 of the circle is shaded in.

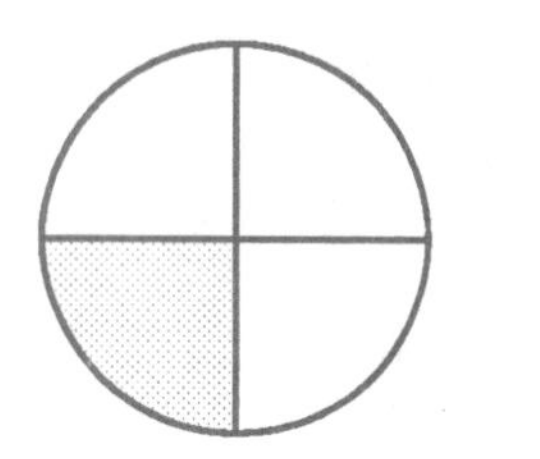

$\frac{1}{4}$ — 1: number of parts that are shaded; 4: number of equal parts in the whole

In a fraction, the top number is called the *numerator*. The bottom number is called the *denominator*.

$\frac{1}{4}$ — 1: Numerator; 4: Denominator

Fractions do not always refer to pictures. They can also refer to situations. For example, if Juanita worked 3 weeks out of a month, what part of the month did she work?

Put the part (3 weeks) on top. Put the whole (4 weeks) on the bottom.

$\frac{3}{4}$ — 3: Part; 4: Whole

Juanita worked $\frac{3}{4}$ of the month.

Exercise 1

Write a fraction for the shaded part of each circle.

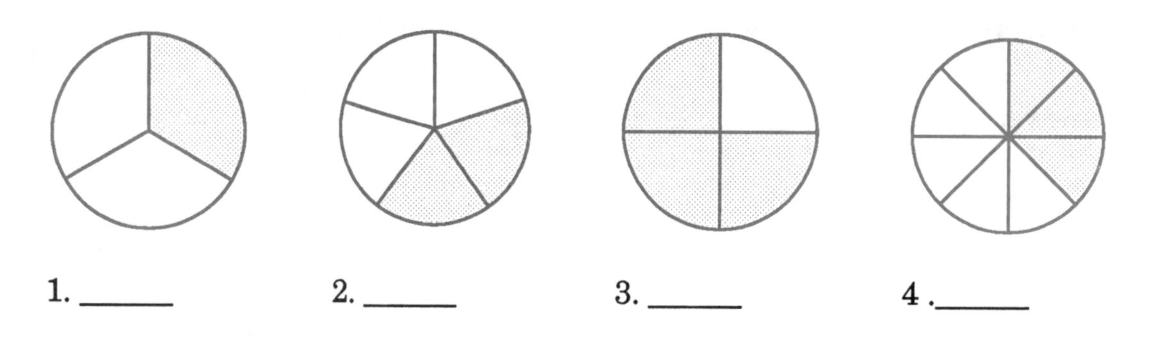

1. ______ 2. ______ 3. ______ 4. ______

Answers begin on page 137

Exercise 2

Look at the circles in Exercise 1 again. Write a fraction for the unshaded part of each circle.

1. ______ 2. ______ 3. ______ 4. ______

Answers begin on page 137.

Exercise 3

Answer each question using a fraction.

1. What part of a year is 5 months? ______________

2. What part of a day is 11 hours? ______________

3. What part of an hour is 30 minutes? ______________

4. What part of a week is 2 days? ______________

5. What part of a dollar is 25 cents? ______________

Answers begin on page 137.

REDUCING FRACTIONS

Different fractions can be used to express the same number relationship. The picture below can be described as either 2/4 or 1/2.

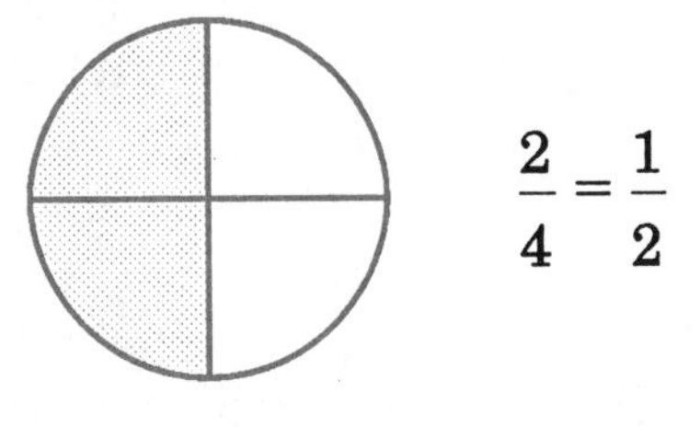

$$\frac{2}{4} = \frac{1}{2}$$

When the lowest possible numbers represent a fraction (in this case 1/2), the fraction is *reduced to its lowest terms*. The example below shows how to reduce 6/10 to lowest terms.

Step 1: Find the highest number that divides into both the numerator and the denominator.

$\frac{6}{10}$ The number 2 goes into both 6 and 10.

Step 2: Divide the numerator and the denominator by 2.

Think: Write:

$\frac{6}{10} \div \frac{2}{2} =$ $\frac{3}{5}$

Exercise 4

Reduce these fractions to lowest terms.

1. 5/15 = **2.** 3/9 = **3.** 4/8 = **4.** 9/12 =

5. 8/24 = **6.** 6/9 = **7.** 10/12 = **8.** 4/16 =

9. 4/20 = **10.** 14/21 = **11.** 16/18 = **12.** 20/24 =

Answers begin on page 137.

CHECKING EQUAL FRACTIONS

When two different fractions are equal (or equivalent), the products of the numbers in opposite "corners" are equal.

$$\frac{2}{5} = \frac{6}{15} \qquad \frac{2}{5} = \frac{6}{15}$$

$$2 \times 15 = 30 \qquad 5 \times 6 = 30$$

Multiplying numbers in opposite corners is called *cross multiplication*. The resulting products are called *cross products*. Use cross multiplication to check reduced fractions. If the cross products are different, then the two fractions are not equivalent.

Exercise 5

Use cross multiplication to check these fractions. If they are equivalent, write "yes." If they are not equivalent, write "no."

1. $\frac{5}{10} = \frac{1}{2}$ ____________

2. $\frac{1}{5} = \frac{5}{15}$ ____________

3. $\frac{3}{12} = \frac{1}{3}$ ____________

4. $\frac{6}{15} = \frac{2}{5}$ ____________

5. $\frac{8}{14} = \frac{4}{7}$ ____________

Check Yourself

Circle the letter of the correct answer for each question. Use the income statement below to answer questions 1 and 2.

Yearly Income Statement
Neighborhood Youth and Family Services

Child Care:	
ages Birth - 5	$120,000
ages 6 - 12	$50,000
Rental Fees (Pool and Gym):	$20,000
Class and Recreation Fees:	$10,000
Grants:	
private	$80,000
state	$80,000
federal	$40,000
TOTAL INCOME:	$400,000

1. Cindy must determine what fraction of the income comes from state grants. The part is ___________

a. $40,000.
b. $80,000.
c. $120,000.
d. $400,000.

2. The whole is ___________

a. $40,000.
b. $80,000.
c. $120,000.
d. $400,000.

3. Colleen works five days a week. Last week she was sick one day. What fraction of the work week did she miss?

a. 5/1
b. 4/5
c. 1/5
d. 1/4

4. To reduce to lowest terms, ___________

a. divide the numerator into the denominator.
b. divide the numerator and the denominator by the same number.
c. multiply the numerator and the denominator by the same number.
d. multiply the numerator by the denominator.

5. Use a fraction to tell ___________

a. what part of the pizza was eaten during lunch.
b. the total yearly income of Neighborhood Youth and Family Services.
c. how much money was contributed through private grants.
d. the yearly expenses of the day care program.

Work Problem

Ted is finishing his yearly report. He must know what part of the total income comes from the child care services that Neighborhood Youth and Family Services provides. Find that fraction for him. Use the income statement on page 135.

Answer: ___________________

Answers to Problem-Solving Practice Questions

DEFINE the problem

- A part of the total income, a fraction
- To express a relationship

PLAN the solution

- Income from the rental of the pool and gym, and total income
- On the income statement
- Income from the rental of the pool and gym, and total income
- Division

SOLVE the problem

- Income from the rental of the pool and gym, and total income
- $\frac{20,000}{360,000}$
- $\frac{20,000}{360,000} = \frac{2}{36} = \frac{1}{18}$
- $20,000 \times 18 = 360,000$
  $1 \times 360,000 = 360,000$
- $\frac{1}{8}$

CHECK the solution

- The defined purpose was accomplished; 1/18 expresses a part.
- The solution to the work problem is reasonable; 1/18 is a fraction less than one.

The income from the rental of the pool and gym is 1/18 of the total income.

Answers to Skills Practice Problems

Exercise 1

1. 1/3 **2.** 2/5 **3.** 3/4 **4.** 3/8

Exercise 2

1. 2/3 **2.** 3/5 **3.** 1/4 **4.** 5/8

Exercise 3

1. 5/12 **2.** 11/24 **3.** 30/60 or 1/2 **4.** 2/7 **5.** 25/100 or 1/4

Exercise 4

1. 1/3 **2.** 1/3 **3.** 1/2 **4.** 3/4 **5.** 1/3 **6.** 2/3

7. 5/6 **8.** 1/4 **9.** 1/5 **10.** 2/3 **11.** 8/9 **12.** 5/6

LESSON 6

Expressing Relationships as Ratios

Some workplace problems require that the relationship between two numbers be expressed as a *ratio*. In some businesses, sales of products are compared using ratios. Sometimes the number of factory workers is compared to the number of supervisors using ratios. At Neighborhood Youth and Family Services, ratios are used in the child care programs to compare the number of children to the number of teachers.

Cindy Spears' solution to the following work problem is shown as an example.

Work Problem

Cindy is responsible for answering the phone at Neighborhood Youth and Family Services. She receives phone calls from parents interested in the child care programs. Early in March, a parent called and said, "I have a two-year-old toddler and am interested in your child care program. What is your child-to-teacher ratio?"

The ratio changes from month to month according to the number of children enrolled in each program. Neighborhood Youth and Family Services maintains an enrollment record, as shown below. The record shows how many children were enrolled in each program during each month. The bottom row tells how many teachers work in each program. Cindy uses the enrollment record to answer parents' questions about child-to-teacher ratios.

ENROLLMENT RECORD
Neighborhood Youth and Family Services

	PROGRAM:				
MONTH:	**Infant**	**Toddler**	**Preschool**	**Elementary**	**TOTALS**
January	8	16	17	37	78
February	6	15	16	36	73
Teachers	2	3	2	4	11

Cindy saw that 15 children were enrolled in the toddler program in February. The program had 3 teachers. She set up the ratio and reduced it to its lowest terms. Determine the equivalent ratio and write your answer below.

$$\frac{\text{Children}}{\text{Teachers}} \quad \frac{15}{3} = \underline{\qquad}$$

Cindy told the parent that the child-to-teacher ratio for February was 5 to 1. Students who did not obtain this answer may need more practice determining ratios. The Skills Practice section on page 146 provides assistance and practice with ratios.

Doing Math to Express Relationships

Cindy knew that determining a ratio meant expressing a relationship. From her plan, she knew she needed the number of children enrolled in February and the number of teachers in the toddler program. She found the data on the enrollment record and planned to divide. Then Cindy solved the problem by following five steps.

1. Select the relevant data.

Cindy used the most recent data available, the enrollment for February. She selected the number of children and the number of teachers.

2. Set up the calculation.

Cindy was asked to compare the number of children to the number of teachers, so she put the number of children (15) on top. She put the number of teachers (3) on the bottom.

$$\frac{15}{3}$$

3. Do the calculation.

Ratios are always expressed in lowest terms, so Cindy reduced the ratio.

$$\frac{15}{3} = \frac{5}{1}$$

4. Check the accuracy of the answer.

Cindy used cross multiplication to check her answer.

$15 \times 1 = 15$

$13 \times 5 = 15$

5. Express the relationship in desired terms.

5 to 1

Now an entire problem will be completed.

Work Problem

Later that same day, another parent called Cindy and said, "I would like to enroll my third grader in your after-school program for elementary students. What is your ratio of children to teachers?"

DEFINE the problem

- What is the expected outcome?

A ratio which compares children to teachers

Note: Cindy must determine how many children to teachers there are. A ratio is a comparison which often uses the word "to."

- What is the purpose?

To express a relationship

Note: When the expected outcome is a ratio, the purpose is to express a relationship.

PLAN the solution

- What data are needed? *The number of elementary students enrolled in February and the number of teachers in the elementary programs on the enrollment record*

- Where can the data be found? *On the enrollment record*

- What is already known? *The number of elementary students enrolled in February and the number of teachers in the elementary program*

- Which operations should be used? *Division*

SOLVE the problem

- Select the relevant data. *The number of elementary students enrolled in February and the number of teachers in the elementary program*

- Set up the calculation.

$$\frac{36}{4}$$

- Do the calculation.

$$\frac{36}{4} = \frac{9}{1}$$

- Check the accuracy of the answer.

$36 \times 1 = 36$

$9 \times 4 = 36$

- Express the relationship in desired terms.

9 to 1

CHECK the solution

Make sure the calculated solution solves the work problem.

- Was the defined purpose accomplished?

Yes, 9 to 1 expresses the relationship of children to teachers

- Is the solution to the work problem reasonable?

Yes, because the number of children should be larger than the number of teachers. 9 is larger than 1.

Note: If a larger number is being compared to a smaller number, then the ratio reduced to lowest terms should also compare a larger number to a smaller number.

The ratio of children to teachers is 9 to 1. Another way to say this is, "There is one teacher for every nine children."

Problem-Solving Practice

Use the DEFINE, PLAN, SOLVE, and CHECK steps to solve the next problem.

Work Problem

Cindy is responsible for answering phone calls at Neighborhood Youth and Family Services. On March 2, a caller says, "I have a preschooler I would like to enroll in your program. What is your child-to-teacher ratio?"

DEFINE the problem

- What is the expected outcome?
- What is the purpose?

PLAN the solution

- What data are needed?
- Where can the data be found?
- What is already known?
- Which operations should be used?

SOLVE the problem

- Select the relevant data.

- Set up the calculation.

- Do the calculation.

- Check the accuracy of the answer.

- Express the relationship in desired terms. ____________________

CHECK the solution

Make sure the calculated solution solves the work problem.

- Was the defined purpose accomplished?

- Is the solution to the work problem reasonable?

- What is the ratio of children to teachers?

Answers to Problem-Solving Practice questions appear on page 151.

On Your Own

Solve the work problems below for added practice. Remember to DEFINE the problem, PLAN the solution, SOLVE the problem, and CHECK the solution.

Work Problem A

On May 1, a parent calls and asks, "How many children do you have in the infant program to each teacher? I have a newborn baby and I want to leave her where she will get lots of attention." Use the enrollment record below to answer the parent's question.

Answer: ____________________

ENROLLMENT RECORD Neighborhood Youth and Family Services					
	PROGRAM:				
MONTH:	Infant	Toddler	Preschool	Elementary	TOTALS
January	8	16	17	37	78
February	6	15	16	36	73
March	7	16	17	38	78
April	8	18	16	40	82
Teachers	2	3	2	4	11

Work Problem B

On May 2, a caller says, "I am calling to inquire about after-school child care for elementary school children. What is your child-to-teacher ratio?" Use the enrollment record above to answer the question.

Answer: ____________________

Work Problem C

Early in May, a parent stops at your desk. He says, "I have a son who is old enough to move from your infant program into your toddler program. How many children to teachers are there in the toddler program?" Use the previous enrollment record to solve the problem.

Answer: ____________________

Work Problem D

All licensed child care centers must meet state requirements for child-to-teacher ratios. Ted must report the ratios at Neighborhood Youth and Family Services to the state. Ted needs to know the child-to-teacher ratio in the toddler program for June. Use the enrollment record below to obtain the answer for Ted.

Answer: ____________________

ENROLLMENT RECORD Neighborhood Youth and Family Services					
	PROGRAM:				
MONTH:	Infant	Toddler	Preschool	Elementary	TOTALS
January	8	16	17	37	78
February	6	15	16	36	73
March	7	16	17	38	78
April	8	18	16	40	82
May	7	19	17	39	82
June	8	21	18	36	83
Teachers	2	3	2	4	11

Work Problem E

Ted must report on the child-to-teacher ratio for the preschool program. He says, "Please determine the number of preschoolers to teachers during June." Use the enrollment record from problem D.

Answer: ____________________

Skills Practice: Ratios

UNDERSTANDING RATIOS

Fractions and ratios both show comparisons and can be written in the same way. However, a *fraction* tells what part of a whole is represented. It compares amounts of the *same kind of units*. A *ratio* compares one kind of unit to another.

For example, computers and printers are two different kinds of units. An office worker could use a ratio to compare them. For the picture below, the office worker could say, "The ratio of computers to printers is 8 to 2."

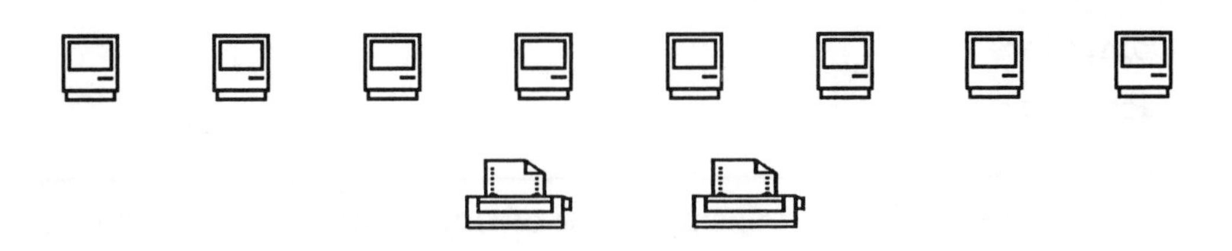

The office worker could write the ratio in one of three ways:

with the word *to*	with a colon (:)	as a fraction
8 to 2	8:2	$\frac{8}{2}$

As fractions, ratios should be reduced to lowest terms.

$$\frac{8}{2} = \frac{4}{1}$$

Expressing the ratio as 4 to 1 does not change the total. It simply makes the comparison easier to understand. Also, the order of the numbers should be the same as the order of the words when expressing ratios. The ratio of *computers to printers* is 4 to 1. The ratio of *printers to computers* is 1 to 4.

Exercise 1

Write a ratio to express each relationship. Remember to reduce the ratio to lowest terms. Use a colon (:).

Example: 10 sales in 2 weeks ______*5:1*______

1. $1 for 3 rolls of tape ______________
2. 95 words a minute ______________
3. 12 pens to a box ______________
4. 10 boxes of paper to 2 weeks ______________
5. 4 hours of sick time for 80 hours of work ______________

Answers begin on page 151.

PROPORTIONS

A proportion is an expression which shows that two ratios are equal. Here is the proportion used in a previous example:

$$\frac{\text{Computers}}{\text{Printers}} \quad \frac{8}{2} = \frac{4}{1}$$

Proportions can be used to find a missing piece of information. For example, suppose a clerk is ordering paper clips priced at 5 boxes for $2. A proportion can be used to find the cost for 15 boxes, as shown below.

Step 1: Set up the proportion.

	Think:		Write:
Boxes of Paper Clips	5	=	15
Cost	2		?

Thinking about the proportion in words will help. The numbers 5 and 15 stand for boxes of paper clips, and those numbers are placed on top. The number 2 stands for the cost and is placed on the bottom.

Step 2: Cross multiply. Multiply the two numbers in opposite "corners."

$$\frac{5}{2} = \frac{15}{?} \qquad 2 \times 15 = 30$$

Step 3: Divide 30 by 5 to find the missing information.

$$\frac{5}{2} = \frac{15}{?} \qquad 30 \div 5 = ?$$

Step 4: Fill in the answer.

$$\frac{5}{2} = \frac{15}{6}$$

Fifteen boxes of paper clips cost $6.

Exercise 2

Find the missing number in each of the following proportions.

1. $\frac{2}{6} = \frac{?}{9}$

2. $\frac{3}{12} = \frac{?}{20}$

3. $\frac{2}{6} = \frac{3}{?}$

4. $\frac{3}{15} = \frac{4}{?}$

5. $\frac{?}{10} = \frac{5}{25}$

Check Yourself

Circle the letter of each correct answer.

1. A ratio is a comparison of ___________
   a. the part to the whole.
   b. the same kind of units.
   c. different units.
   d. two equivalent fractions.

2. Check if two ratios are equal by ___________
   a. multiplying the first numerator by the second numerator.
   b. multiplying the first denominator by the second denominator.
   c. cross multiplying.
   d. cross dividing.

3. There are two people for every typewriter in an office. The ratio of people to typewriters could be written as ___________
   a. all of the below.
   b. 2/1.
   c. 2:1.
   d. 2 to 1.

4. An expression showing two equal ratios is a __________

a. fraction.

b. decimal.

c. percent.

d. proportion.

5. Ramon Manus is a clerk typist. Last week, he typed 20 documents in 8 hours. What is his ratio of documents to hours?

a. 8:1

b. 8:20

c. 5:2

d. 5:4

Work Problem

Ted is working on a report to the state department. He needs to compare the number of children to the number of teachers in the elementary program for the month of August. Use the enrollment record below to find the ratio needed.

Answer: ________________

ENROLLMENT RECORD
Neighborhood Youth and Family Services

	PROGRAM:				
MONTH:	**Infant**	**Toddler**	**Preschool**	**Elementary**	**TOTAL**
January	8	16	17	37	78
February	6	15	16	36	73
March	7	16	17	38	78
April	8	18	16	40	82
May	7	19	17	39	82
June	8	21	18	36	83
July	7	20	19	33	79
August	6	21	20	32	79
Teachers	2	3	2	4	11

Answers to Problem-Solving Practice Questions

DEFINE the problem

- The ratio of children to teachers in the preschool program
- To express a relationship

PLAN the solution

- The number of children in the preschool program in February and the number of teachers
- On the enrollment record
- The number of children in the preschool program in February and the number of teachers
- Division

SOLVE the problem

- The number of children in the preschool program in February and the number of teachers
- $\frac{16}{2}$
- $\frac{16}{2} = \frac{8}{1}$
- $16 \times 1 = 16$
  $2 \times 8 = 16$
- 8 to 1

CHECK the solution

- The defined purpose was accomplished. The ratio, 8 to 1, expresses the relationship of children to teachers.
- The solution to the work problem is reasonable. One would expect the number of children to be larger than the number of teachers. Eight is larger than one.

There were 8 children to 1 teacher.

Answers to Skills Practice Problems

Exercise 1

1. 1:3 **2.** 95:1 **3.** 12:1 **4.** 5:1 **5.** 1:20

LESSON 7

Expressing Relationships as Percents

Relationships between numbers can be expressed as *percents*. Percents are used widely in business. One out of every twelve employees has been with Bostic Company for fifteen years or less. What percent does this represent? A marketing survey team wishes to know the percent of customers who prefer a new product. A time management team wishes to know the percent of time that office support personnel spend on the telephone. All of these situations require expressing the relationship between numbers as a percent.

Colleen Crosby's solution to the following work problem is shown as an example.

Work Problem

Colleen Crosby is the coordinator for the child care program that serves children age five and under. She was preparing her annual report and needed to know what percent salaries represented of the total expenses for her program.

EXPENSE SUMMARY
Programs at Neighborhood Youth and Family Services

	AGE GROUP:				
EXPENSE:	**Ages B-5**	**Ages 6-12**	**Ages 13-18**	**Adult**	**TOTAL**
Salaries	128,000	50,160	24,480	19,500	222,140
Supplies	12,800	6,600	2,160	1,200	22,760
Building and Maintenance	19,200	9,240	9,360	9,300	47,100
TOTAL	160,000	66,000	36,000	30,000	292,000

Neighborhood Youth and Family Services serves four age groups. The expense summary above itemizes the expenses for each age group. The expenses fall into three categories: salaries, supplies, and building and maintenance.

Salaries for Colleen's program were $128,000 and total expenses were $160,000. Colleen set up the fraction below and converted it to a percent. What percent does the fraction equal?

$$\frac{128,000}{160,000} =$$

Colleen found that salaries were 80% of the expenses. Students who obtained a different answer may need more practice determining percents. The section beginning on page 161 provides assistance and practice with percents.

Doing Math to Express Relationships

Colleen needed to express the relationship between salaries and total expenses as a percent. She planned to locate the amount for salaries and the total expenses on the expense summary and divide. Then she solved the problem, following the five steps below.

1. Select the relevant data.

The expense for salaries and the total expenses

2. Set up the calculation.

Colleen was expressing the relationship between the salaries and the total expenses. She set up the numbers as a fraction, putting the part (salaries) on top and the whole (total expenses) on the bottom.

$$\frac{128,000}{160,000}$$

3. Do the calculation.

Reduce to lowest terms

$$\frac{128,000}{160,000} = \frac{4}{5}$$

Divide to determine the equivalent decimal

$$5\overline{)4.0} = .8$$

$$\begin{array}{r} 4\,0 \\ \hline 0 \end{array}$$

4. **Check the accuracy of the answer.**

Colleen checked her answer by multiplying 8 by 160,000.

$$\begin{array}{r} 160{,}000 \\ \times\ \ .8 \\ \hline 128{,}000 \end{array}$$

5. **Express the relationship in desired terms.**

Convert the decimal to a percent by moving the decimal point 2 places to the right: .8 = 80%

Salaries were 80% of the expenses.

Now an entire problem will be completed by Peter Gonzales.

Work Problem

Pete is the coordinator for the junior and senior high program, ages 13-18. He was preparing his annual report, just as Colleen was. Pete needed to know what percent building and maintenance expense was of his total expenses. He used the same expense summary as Colleen.

DEFINE the problem

- What is the expected outcome? *A percent of the total income*
- What is the purpose? *To express a relationship*

Note: When the expected outcome is a percent, the purpose is to express a relationship.

PLAN the solution

- What data are needed? *The expenses for building and maintenance and the total expenses*

- Where can the data be found? *On the expense summary*

- What is already known? *The expenses for building and maintenance and the total expenses*

- Which operations should be used? *Division*

SOLVE the problem

- Select the relevant data. *Building and maintenance expenses and total expenses*

- Set up the calculation.

$$\frac{24{,}480}{36{,}000}$$

- Do the calculation.

$$\frac{24{,}480}{36{,}000} = \frac{51}{75}$$

$$\begin{array}{r} .68 \\ 75\overline{)51.00} \\ \underline{450} \\ 600 \\ \underline{600} \\ 0 \end{array}$$

- Check the accuracy of the answer.

```
   36,000
  ×   .68
   288000
  216000
24,480.00
```

- Express the relationship in desired terms.

  .68 = 68%

CHECK the solution

Make sure the calculation solves the work problem.

- Was the defined purpose accomplished?

  Yes, 68% expresses a relationship

- Is the solution to the work problem reasonable?

  Yes, 68% is reasonable because it is less than 100%

Note: One hundred percent (100%) expresses a whole. When a part (in this case, 24,480) is compared to the whole (36,000), then the percent is smaller than 100 (e.g. 68%). When the part is larger than the whole, the percent is greater than 100.

Problem-Solving Practice

Use the DEFINE, PLAN, SOLVE, and CHECK steps to solve the next problem.

Work Problem

Beverly Wong is the coordinator of the program for six-to-twelve-year-old children. She, too, is working on her annual report. You are an assistant in Beverly's program. She asks, "What percent of the total expenses for the six-to-twelve-year-old children related to salaries?" Use the expense summary at the beginning of this lesson to solve this problem.

DEFINE the problem

- What is the expected outcome?

- What is the purpose?

PLAN the solution

- What data are needed?

- Where can the data be found?

- What is already known?

- Which operations should be used?

SOLVE the problem

- Select the relevant data.

- Set up the calculation.

- Do the calculation.

- Check the accuracy of the answer.

- Express the relationship in desired terms. ____________________

CHECK the solution

Make sure the calculation solves the work problem.

- Was the defined purpose accomplished? ____________________
____
____

- **Is the solution to the work problem reasonable?** ____________________

- **What percent of the expenses for Beverly's program went for salaries?** ____________________

Answers to Problem-Solving Practice questions appear on page 168.

On Your Own

Solve the work problems below. Remember to DEFINE the problem, PLAN the solution, SOLVE the problem, and CHECK the solution. Make sure the solution solves the work problem.

Work Problem A

Vern Washington is the coordinator for the adult program. He says, "Please determine the percent of my expenses that related to building and maintenance."

Answer: ____________________

EXPENSE SUMMARY
Programs at Neighborhood Youth and Family Services

	AGE GROUP:				
EXPENSE:	**Ages B-5**	**Ages 6-12**	**Ages 13-18**	**Adult**	**TOTAL**
Salaries	128,000	50,160	24,480	19,500	222,140
Supplies	12,800	6,600	2,160	1,200	22,760
Building and Maintenance	19,200	9,240	9,360	9,300	47,100
TOTAL	160,000	66,000	36,000	30,000	292,000

Work Problem B

Vern also wants some other information. He asks, "What percent of the expenses related to supplies?"

Answer: ____________________

Work Problem C

Ted Urbanski, the director, is preparing a report for the board of trustees. He says, "Find out what percent of the total salary expense is paid to the staff who work in the program for children under five."

Answer: ____________________

EXPENSE SUMMARY
Programs at Neighborhood Youth and Family Services

	AGE GROUP:				
EXPENSE:	**Ages B-5**	**Ages 6-12**	**Ages 13-18**	**Adult**	**TOTAL**
Salaries	131,100	52,900	25,300	20,700	230,000
Supplies	13,000	4,000	2,000	1,000	20,000
Building and Maintenance	20,000	10,500	10,000	9,500	50,000
TOTAL	164,100	67,400	37,300	31,200	300,000

Work Problem D

Ted is still working on the report. Ted says, "We seem to spend the least amount of money on the adults. What percent of the total salary expense goes to those who work with adults?"

Answer: ____________________

Work Problem E

Ted would like to compare what is spent on the programs for children and youth (ages up to 18) and what is spent on the adults. He asks, "What percent of the salary expense goes for all the youth programs combined?"

Answer: ____________________

Skills Practice: Percents

UNDERSTANDING PERCENTS

Remember that fractions are used to express the relationship between a part and a whole. If 3 out of 8 people in an office take the bus to work, then 3/8 of the people in the office are bus riders.

$$\frac{\text{Part}}{\text{Whole}} \quad \frac{3}{8}$$

Percents are also used to express the relationship between a part and a whole. When using percents, the whole is divided into 100 parts. Percent means "per 100" or "out of 100." If 34 out of 100 people in a company take the bus to work, then 34% of the people at the company ride the bus.

$$\frac{\text{Part}}{\text{Per 100}} \quad \frac{34}{100} = 34\%$$

The diagram below shows 34% in picture form; 34 of the 100 squares are shaded in.

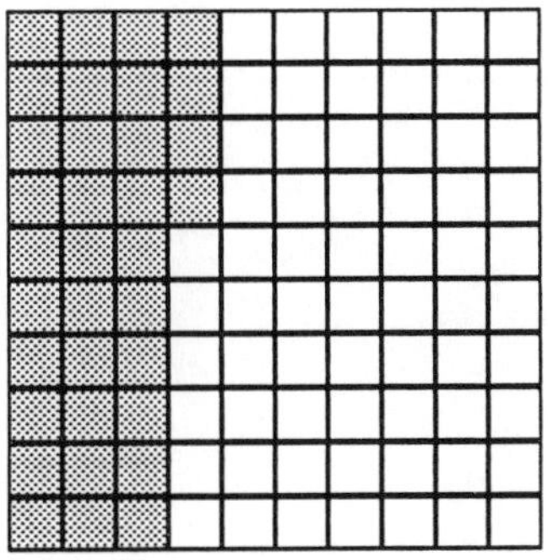

Exercise 1

Find the percents described in these situations.

1. In a department store, 100 customers were shopping. Of these customers, 37 arrived before 9 A.M. What percent of the customers arrived before 9 A.M.?

Answer: ____________________

2. Bankston Manufacturing Co. manufactures a very delicate part for electronic equipment. A study shows that 1 part per 100 manufactured will be defective. What percent of the parts are likely to be defective?

Answer: ____________________

3. A data entry clerk averages 2 errors for every 100 keystrokes. What percent of error results?

Answer: ____________________

4. Munford Educational Data Systems normally has 100 employees in the work force. History shows that an average of 7 employees will move or be transferred each year. What percent of employees will move or be transferred each year?

Answer: ____________________

5. The office supply store had 100 bottles of laser printer toner in stock. A month later, they had 68 bottles in stock. (a) What percent of the inventory remains unsold? (b) What percent of the inventory has been sold?

Answers:

(a) ____________________

(b) ____________________

Answers begin on page 169.

CONVERTING DECIMALS AND FRACTIONS TO PERCENTS

Although decimals and fractions express a relationship, they are often converted to percents in order to express the relationship in different terms. To convert a decimal to a percent, move the decimal point two places to the right and add a percent sign. For example, .17 becomes 17%.

Here are some other examples.

0.05 becomes 5%

0.45 becomes 45%

0.125 becomes 12.5%

1.23 becomes 123%

$.66\frac{2}{3}$ becomes $66\frac{2}{3}\%$

Converting a fraction to a percent is easy when the denominator in the fraction is 100. Percent means parts per 100, so when the denominator is 100, the numerator is the percent. For example, 18/100 = 18%.

Most of the time, though, the denominator is not 100. To convert a fraction to a percent, first convert the fraction to a decimal. For example, do the following to convert 5/8 to a percent:

Step 1: Divide the numerator by the denominator to convert the fraction to a decimal.

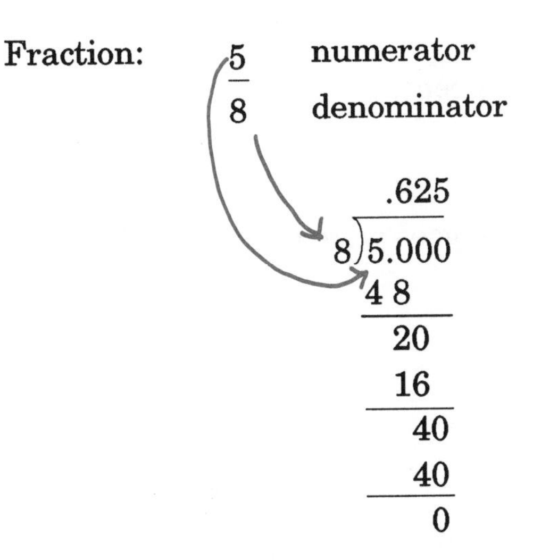

Step 2: Convert the decimal to a percent by moving the decimal point 2 places to the right.

Decimal	Percent
.625	62.5%

When the fraction has a large numerator or denominator, reduce the fraction to lowest terms before dividing. Smaller numbers make division easier. Suppose that 36 out of 48 employees were union members. To determine what percent were union members, follow the steps shown below.

Step 1: Set up the fraction.

$$\frac{36}{48}$$

Step 2: Reduce the fraction to lowest terms.

$$\frac{36}{48} = \frac{3}{4} \quad \begin{matrix}\text{(Divide 36 by 12.)} \\ \text{(Divide 48 by 12.)}\end{matrix}$$

Step 3: Divide the numerator by the denominator to get a decimal.

$$\begin{array}{r} .75 \\ 4\overline{)3.00} \\ \underline{2\,8} \\ 20 \\ \underline{20} \\ 0 \end{array}$$

Step 4: Convert the decimal to a percent.

.75 = 75%

Seventy-five percent (75%) of the employees were union members.

Exercise 2

Convert each of the following to a percent.

1. 0.25 =

2. 0.57 =

3. 0.89 =

4. 0.07 =

5. 0.075 =

6. 0.1875 =

7. 0.87285 =

8. 0.03 1/3 =

9. 1.25 =

10. 1 =

Answers begin on page 169.

Exercise 3

Convert each of the following fractions to decimals and percents. Remember to reduce the fractions to lowest terms when possible.

	Fraction	Decimal	Percent
1.	33/100		
2.	7/10		
3.	1/2		
4.	3/5		
5.	1/4		
6.	17/20		
7.	11/25		
8.	21/50		
9.	27/30		
10.	12/15		

Answers begin on page 169.

Exercise 4

Convert each of the following decimals or fractions to percents.

1. 0.49 =

2. 0.68 =

3. 0.33 1/3 =

4. 0.155 =

5. 23/100 =

6. 3/10 =

7. 1/5 =

8. 1/25 =

9. 15/20 =

10. 75/150 =

Check Yourself

EXPENSE SUMMARY Programs at Neighborhood Youth and Family Services					
	AGE GROUP:				
EXPENSE:	Ages B-5	Ages 6-12	Ages 13-18	Adult	TOTAL
Salaries	131,100	52,900	25,300	20,700	230,000
Supplies	13,000	7,000	2,000	1,000	23,000
Building and Maintenance	20,000	10,500	10,000	9,500	50,000
TOTAL	164,100	70,400	37,300	31,200	303,000

1. What information can be determined from the expense summary above?

a. How much money Neighborhood Youth and Family Services spent on buildings and maintenance for four programs

b. How much income Neighborhood Youth and Family Services received from supplies for four programs

c. Neighborhood Youth and Family Services' total expenses

d. Both a and c above

2. To determine the percent of the salary expense that was budgeted for salaries for the high school program, set up the problem this way:
   a. 25,300/37,300.
   b. 37,300/25,300.
   c. 25,300/230,000.
   d. 230,000/25,300

3. To solve the problem above, follow three steps in this order:
   a. (1) convert the fraction to a decimal, (2) reduce the fraction, (3) convert the decimal to a percent.
   b. (1) reduce the fraction, (2) convert the fraction to a decimal, (3) convert the decimal to a percent.
   c. (1) convert the decimal to a percent, (2) convert the fraction to a decimal, (3) reduce the fraction.
   d. (1) convert the fraction to a decimal, (2) convert the decimal to a percent, (3) reduce the fraction.

4. One hundred employees work at Teutsch Manufacturing Company. Forty-two of the employees are female. What percent of the employees are female?
   a. 42%
   b. 58%
   c. 100%
   d. 142%

5. What is 13/52 expressed as a percent?
   a. 4%
   b. 13%
   c. 25%
   d. 52%

Work Problem

Ted is preparing his report to the board of trustees. He asks, "What percent of the total supply expense was spent on the program for children five and under?" Use the expense summary on page 166.

Answer: ____________________

Answers to Problem-Solving Practice Questions

DEFINE the problem

- A percent of the total income
- To express a relationship

PLAN the solution

- Salary expenses and total expenses
- On the expense summary
- Salary expenses and total expenses
- Division

SOLVE the problem

- Salary expenses and total expenses
- $\frac{50,160}{66,000}$
- $\frac{50,160}{66,000} = \frac{19}{25}$
- $$\begin{array}{r} .76 \\ 25\overline{)19.0} \\ \underline{17\,5} \\ 1\,50 \\ \underline{1\,50} \\ 0 \end{array}$$
- $$\begin{array}{r} 66,000 \\ \underline{\times .76} \\ 3960\,00 \\ \underline{46200\,0} \\ 50160.00 \end{array}$$
- .76 = 76%

CHECK the solution

- The defined purpose for calculating was accomplished; 76% expresses a relationship.
- The solution to the work problem is reasonable because 76% is less than 100%.
- 76% of the expenses were for salaries.

Answers to Skills Practice Problems

Exercise 1

1. 37% **2.** 1% **3.** 2% **4.** 7% **5(a).** 68% **5(b).** 32%

Exercise 2

1. 25% **2.** 57% **3.** 89% **4.** 7% **5.** 7.5%

6. 18.75% **7.** 87.285% **8.** 3 1/3% **9.** 125% **10.** 100%

Exercise 3

1. .33, 33% **2.** .7, 70% **3.** .5, 50% **4.** .6, 60% **5.** .25, 25%

6. .85, 85% **7.** .44, 44% **8.** .42, 42% **9.** .9, 90% **10.** .8, 80%

LESSON 8

Finding Amounts With Percents

Percents are used to determine discounts, pay increases, sales commissions, taxes, and interest. In many work problems, office workers use percents to determine a missing amount. Although the calculation requires the use of a percent, the purpose is to find an amount, so the strategy steps used in this lesson are the same as those in Unit I.

Darren Simms' solution to the following work problem is shown as an example.

Work Problem

Darren is responsible for calculating the payroll at Neighborhood Youth and Family Services. He records the calculations in the *payroll ledger*, a record book for payroll. Each page summarizes employees' earnings, deductions, and net pay for a particular pay period. The report below covers the pay period of June 17 to June 30, 199–.

Ted sent Darren a memo which said, "Colleen Crosby will be receiving a 5% cost-of-living raise, beginning July 1. Please tell me the dollar amount of Colleen's raise."

PAYROLL SUMMARY Neighborhood Youth and Family Services							
PAY PERIOD:	**June 17 - June 30, 199–**						
NAME	**HOURS**	**EARNINGS**	**DEDUCTIONS**				**NET PAY**
			FICA	**Fed tax**	**State tax**	**Total**	
Kathy Barnes	40	396.00	25.74	43.56	15.84	85.14	310.86
Colleen Crosby	80	976.00	63.44	107.36	39.04	209.84	766.16
Juanita Estevez	40	292.00	18.98	32.12	11.68	62.78	229.22
Peter Gonzales	80	952.00	61.88	104.72	38.08	204.68	747.32
Philip Leighton	40	308.00	20.02	33.88	12.32	66.22	241.78
Sam May	30	162.00	10.53	17.82	6.48	34.83	127.17
Pat McDonold	40	412.00	26.78	45.32	16.48	88.58	323.42
Darren Simms	80	735.00	47.84	80.98	29.44	158.26	576.74
Cindy Spears	80	648.00	42.12	71.28	25.92	139.92	508.68
Ted Urbanski	80	1200.00	78.00	132.00	48.00	258.00	942.00
Vern Washington	40	488.00	31.72	53.68	19.52	104.92	383.08
Beverly Wong	80	952.00	61.88	104.72	38.08	204.68	747.32

Darren had to find 5% of $976, which was Colleen's salary. Use the space below to determine the amount of Colleen's pay raise.

Darren converted 5% to .05 and multiplied by 976. Colleen's salary increased by $48.80. Students who obtained a different answer may need more practice determining amounts with percents. The section beginning on page 178 provides assistance and practice with finding amounts by using percents.

Doing Math to Find Amounts

Darren had to determine Colleen's pay increase, which is an amount. He located her earnings in the ledger and planned to multiply by .05 (or 5%). Then Darren solved the problem by following four steps.

1. **Select the relevant data.** *Colleen's earnings and the percent of increase*

2. **Set up the calculation.**

$$\begin{array}{r} \$976 \\ \times\ .05 \\ \hline \end{array}$$

Note: Before Darren set up the calculation, he converted 5% to .05.

3. **Do the calculation.**

$$\begin{array}{r} \$976 \\ \times\ .05 \\ \hline \$48.80 \end{array}$$

4. **Check the accuracy of the answer.**

$$\begin{array}{r} .05 \\ 976\,\overline{)\,48.80} \\ 48\ 80 \\ \hline 0 \end{array}$$

Colleen will be earning an extra $48.80 per pay period.

Now an entire problem will be completed.

Work Problem

Ted sent Darren a memo that said, "Beginning July 1, Cindy Spears will be receiving a 6% cost-of-living raise. Please determine the dollar amount of her raise."

DEFINE the problem

- What is the expected outcome? *The amount of Cindy's raise*

- What is the purpose? *To find an amount*

PLAN the solution

- What data are needed? *The percent of increase and Cindy's earnings from the last pay period*

- Where can the data be found? *The percent is on the memo. Cindy's earnings are in the payroll ledger.*

- What is already known? *The percent of increase and Cindy's earnings from the last pay period*

- Which operation should be used? *Multiplication*

Note: Darren must determine 6% of $648, which is a percent (6%) of a base (648). Multiply whenever the percent and the base are known.

SOLVE the problem

- Select the relevant data. *The percent of increase and Cindy's earnings from the last pay period*

- Set up the calculation.

$$\begin{array}{r} \$\,648 \\ \times\ .06 \\ \hline \end{array}$$

- Do the calculation.

$38.88

- Check the accuracy of the answer.

$$\begin{array}{r} .06 \\ 648\overline{)38.88} \\ \underline{38\ 88} \\ 0 \end{array}$$

CHECK the solution

Make sure the calculation solves the work problem.

- Was the defined purpose accomplished?

Yes, $38.88 is a dollar amount.

- Is the solution to the work problem reasonable?

Yes, because $38.88 is less than $648.00

Note: This problem requires finding a part of $648. The part is represented by the decimal .06, which is less than one. Whenever multiplying a decimal that is less than one by a whole number, expect the answer to be less than the original whole number.

Problem-Solving Practice

Use the DEFINE, PLAN, SOLVE, and CHECK steps to solve the next problem.

Work Problem

Several people at Neighborhood Youth and Family Services will receive raises on July 1. Ted sends the bookkeeper a memo stating that Philip Leighton will receive a 5% raise. Philip asks the bookkeeper, "How much more will I make each pay period beginning July 1?"

DEFINE the problem

- What is the expected outcome?
- What is the purpose?

PLAN the solution

- What data are needed?
- Where can the data be found?
- What is already known?
- Which operation should be used?

SOLVE the problem

- Select the relevant data.

- **Set up the calculation.**

- **Do the calculation.**

- **Check the accuracy of the answer.**

CHECK the solution

Make sure the calculation solves the work problem.

- **Was the defined purpose accomplished?** ______________________

- **Is the solution to the work problem reasonable?** ______________________

Answers to the Problem-Solving Practice questions appear on pages 188.

On Your Own

Solve the following work problems. Refer to the payroll summary at the beginning of this lesson.

Work Problem A

Sam May, the lifeguard at the pool, calls and says, "Ted said I am getting a 4.5% raise. How much more will I be making per pay period?"

Answer: ____________________

Work Problem B

Ted sent the following message to the bookkeeping department: "Starting July 1, Juanita Estevez gets a 9% raise, 5% for cost of living and 4% for completing her fourth year. Find out how much more Juanita will be earning per pay period."

Answer: ____________________

Work Problem C

Pat McDonald will be receiving a 6% raise on July 1. Find Pat's new gross earnings so the amount can be recorded in the ledger.

Answer: ____________________

Work Problem D

Ted says, "Kathy Barnes' 8% raise includes 5% for cost of living and 3% for working here for two years." Find Kathy's new gross earnings so the amount can be recorded in the ledger.

Answer: ____________________

Work Problem E

A year ago, Darren Simms received a 5% raise. He now earns $735 per pay period. How much was Darren earning per pay period a year ago?

Answer: ____________________

Skills Practice: Amount, Base, and Rate

CONVERTING PERCENTS TO DECIMALS

Before percents can be used in calculations, they must be converted to decimals. To convert a percent to a decimal, move the decimal point two places to the left and remove the percent sign.

37.5% becomes .375

24% becomes .24

4% becomes .04 (Notice that a zero was added.)

66⅔% becomes .66⅔

Exercise 1

Convert each of the following percents to a decimal.

	Percent	Decimal
Example:	20%	0.20 (or 0.2)
1.	30%	
2.	1.5%	
3.	6%	
4.	.9%	
5.	25%	
6.	101%	
7.	49%	
8.	75%	
9.	33⅓%	
10.	17%	

Answers begin on page 188.

UNDERSTANDING AMOUNT, BASE, AND RATE

Work problems with percents consist of three items of information: the *rate* (percent), the *amount*, and the *base*. Identification of each item is necessary to set up the problem and perform the calculation. The following situation gives examples of amount, base, and rate.

John Lee made $1250 per pay period. He received a raise of 10%. Now John earns $125 more each pay period.

Base = $1250	Think of the base as the original amount or the whole. John's whole salary is $1250 per pay period.
Rate = 10%	Often the rate is a percent. Look for the percent sign (%). A rate can also be a decimal or a fraction.
Amount = $125	Think of the amount as the part. The part by which John's salary increased is $125.

THE BASIC FORMULA

Amount, base and rate are related through multiplication. The basic formula which shows this relationship is

$$\text{Amount} = \text{Base} \times \text{Rate}$$

In other words, to determine the amount, multiply the base by the rate.

To determine the base or the rate, simply rearrange the formula. Division is the opposite of multiplication, so

$$\text{Base} = \text{Amount} \div \text{Rate}$$

$$\text{Rate} = \text{Amount} \div \text{Base}$$

The triangle below provides an easy way to remember when to multiply and when to divide.

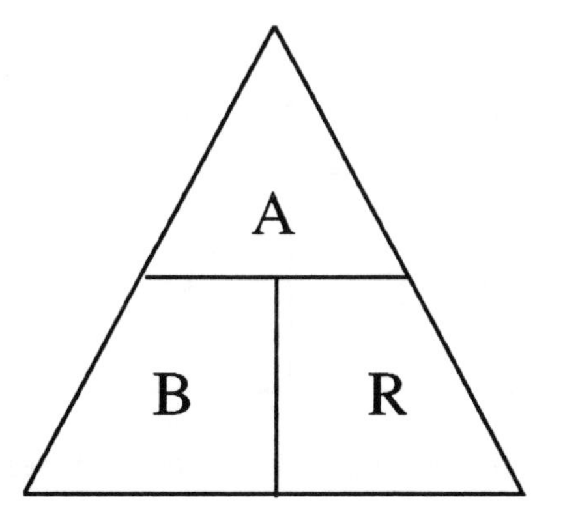

Let the horizontal bar (——) in the triangle represent a division symbol. Let the vertical line (|) represent a multiplication symbol. Cover up the factor you are trying to determine. The relationship of the remaining factors will provide the formula.

For example, to determine the formula for amount, cover up the A. $B|R$ is left showing:

$$\text{Amount} = \text{Base} \times \text{Rate}$$

To determine the formula for the base, cover up the B. A/R is left showing.

$$\text{Base} = \text{Amount} \div \text{Rate}$$

To determine the formula for the rate, cover up the R. A/B is left showing.

$$\text{Rate} = \text{Amount} \div \text{Base}$$

The rest of this lesson will provide instruction and practice in applying the formulas.

FINDING THE AMOUNT

Here is a sample problem in which determining the amount is necessary.

> In the Smith Co., 32 employees work in the office. A recent report showed that 25% of those employees earn more than $30,000 each year. How many of those employees earn more than $30,000?

The rate is 25% and the base is 32. The problem is to find the amount, which is 25% of 32.

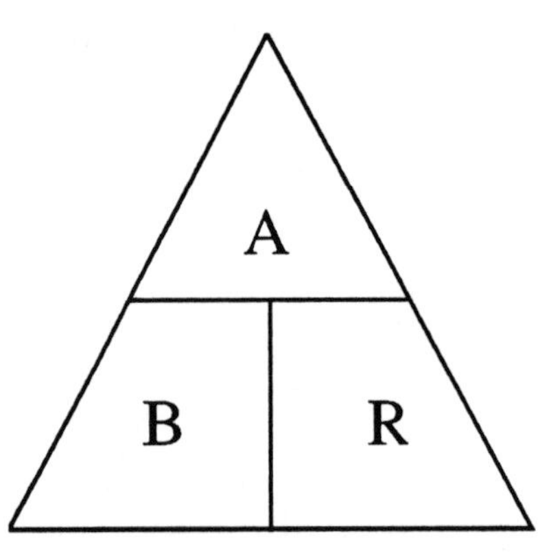

The amount is a part of the base. Multiply whenever determining a part of the base.

Step 1: Change the percent to a decimal rate.

25% becomes .25

Step 2: Multiply the base by the rate.

$$\begin{array}{r} 32 \\ \times\ .25 \\ \hline 160 \\ +\ 64\ \ \\ \hline 8.00 \end{array}$$

Of the 32 employees, 8 earn more than $30,000.

Exercise 2

Find the amount in each of the following problems. Remember to convert the percents to decimals before solving the problems.

1. 50% of 30

2. 32% of 120

3. 7% of 218

4. 3% of $221

5. 15.5% of 400

6. 10.25% of 1,000

7. 25% of $720

8. 35% of $800

Answers begin on page 189.

FINDING THE RATE

Finding the rate is the same as converting a fraction to a percent. This was presented in Lesson 3. Think of the amount as the part and the base as the whole. To find the rate, divide the amount by the base.

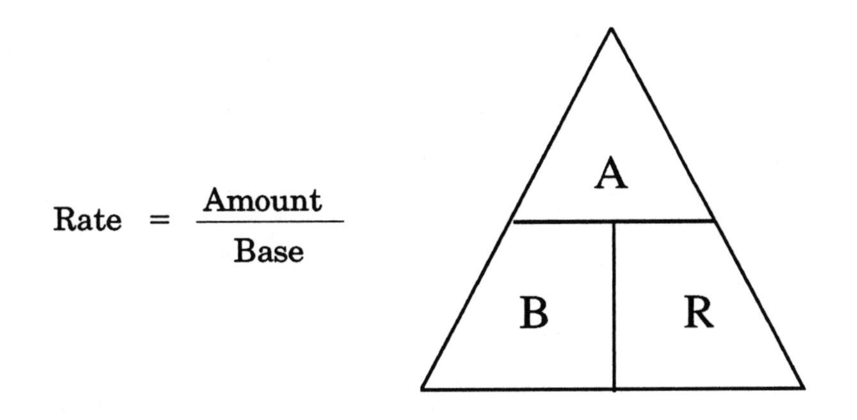

Here is a sample problem.

> Barbara Banks earned $15,000 last year. This year she received a raise of $1,200. Barbara wants to know the percent of her raise.

The amount is 1200 and the base is 15,000. The problem is to find

$$\text{Rate} = \frac{\text{Amount}}{\text{Base}} \quad \text{or} \quad \frac{1200}{15000}$$

Step 1: Divide the amount by the base.

$$\begin{array}{r} .08 \\ 15000\overline{)1200.00} \\ \underline{1200.00} \\ 0 \end{array}$$

Step 2: Convert the decimal rate to a percent by moving the decimal point 2 places to the right.

$$.08 = 8\%$$

Barbara received an 8% raise.

Exercise 3

Find the rate in each of the following problems. An example is provided. Remember to convert the rate to a percent when solving for the answer.

	Amount	Rate	Base
Example:	$93.75	12.5%	$750.00
1.	200	_____	800
2.	30	_____	750
3.	85	_____	850
4.	35	_____	700
5.	14.025	_____	280.5

Answers begin on page 189.

FINDING THE BASE

To find the base, divide the amount by the rate.

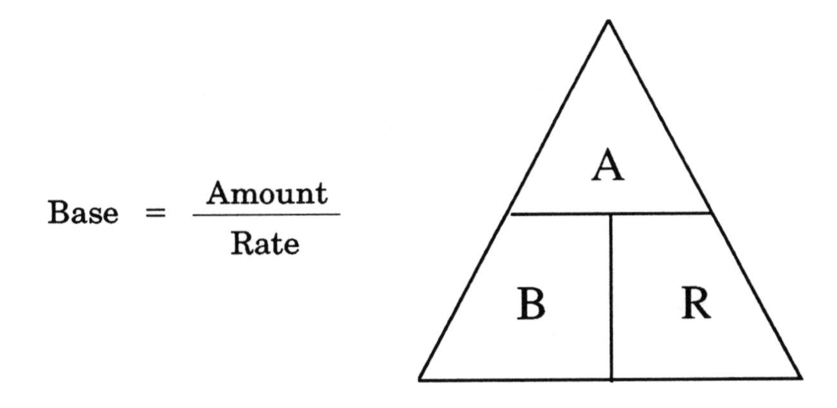

Here is a sample problem.

> Mike Alvarez receives a raise of $24 per week. This represents 6% of his weekly salary. How much was Mike earning per week before the raise?

The amount is 24 and the rate is 6%. The problem is to find

$$\frac{\text{Amount}}{\text{Rate}} \quad \text{or} \quad \frac{24}{06}$$

Step 1: Convert the percent to a decimal rate.

$$6\% = .06$$

Step 2: Divide the amount by the decimal rate.

```
     4.00
.06)24.00
    24
     0 00
```

Note: Move the decimal point 2 places to the right to make .06 a whole number. Do the same with 24.

Mike earned $400 a week before the raise.

Exercise 4

Identify the base, rate and amount in each of the problems below. Use a ? for the information that is not known. Do NOT solve these problems.

Example: Dan Hughes received a raise of 3%. He is now making $45.00 more per pay period. What was he making before his raise?

Base ____?____

Rate ____3%____

Amount ____$45.00____

1. The weekly sales goal for Daryl Quintin was $4,000. During the week, Daryl sold 92% of his goal. How much did he sell?

Base ____________

Rate ____________

Amount ____________

2. The annual sales conference was attended by 320 more people this year than last year. This is a 25% increase. How many people attended last year's conference?

Base ____________

Rate ____________

Amount ____________

3. Last year, Paula Crawford earned $520 per week. This year, she will be making $41.60 more per week. What is the percent of her raise?

Base ____________

Rate ____________

Amount ____________

4. A customer owes $850 on his credit account. He paid 70% of the $850. How much did he pay?

Base ____________

Rate ____________

Amount ____________

5. Anne Carpenter earned $420 last week. County taxes of $8.40 were deducted from her check. What is the tax rate in the county where Anne lives?

Base ____________

Rate ____________

Amount ____________

Answers begin on page 189.

Exercise 5

Find the base for each of the following problems. An example is provided. Remember to convert the percent to a decimal value for the rate.

	Amount	Rate	Base
Example:	$35.63	7%	$509.00
1.	135	15%	________
2.	791	5%	________
3.	375	25%	________
4.	$197.50	12.5%	$________
5.	$69.12	4%	$________

Answers to begin on page 189.

Exercise 6

Supply the missing information for each of the following problems. An example is provided.

	Amount	Rate	Base
Example:	8.415	3%	280.50
1.	________	4.5%	10.8
2.	________	6%	700
3.	600	4%	________
4.	150	______	3,000
5.	$________	12.5%	$664.64

Exercise 7

Solve these problems from Exercise 2.

1. The weekly sales goal for Daryl Quintin was $4,000. During the week, Daryl sold 92% of his goal. How much did he sell? ____________

2. The annual sales conference was attended by 320 more people this year than last year. This is a 25% increase. How many people attended last year's conference? ____________

3. Last year, Paula Crawford earned $520 per week. This year, she will be making $41.60 more per week. What is the percent of her raise? ____________

4. A customer owes $850 on his credit account. He paid 70% of the $850. How much did he pay? ____________

5. Anne Carpenter earned $420 last week. County taxes of $8.40 were deducted from her earnings. What is the tax rate in the county where Anne works? ____________

Check Yourself

Circle the letter of the correct answer.

1. A payroll summary includes ____________
   a. each employee's earnings.
   b. each employee's deductions.
   c. each employee's net pay.
   d. all of the above.

Use the information below to answer questions 2 through 5:

Yvette Brown earned $24,000 last year. On January 1, she received a 6% raise. How much more will Yvette be earning this year?

2. In this problem, the expected outcome is ___________
  a. the dollar amount of Yvette's raise.
  b. the percent of Yvette's raise.
  c. Yvette's earnings last year.
  d. Yvette's new net pay.

3. To solve this problem, ___________
  a. divide 24,000 by .06.
  b. divide .06 by 24,000.
  c. multiply .06 by 24,000.
  d. none of the above.

4. How much more will Yvette be earning?
  a. $144
  b. $1440
  c. $400
  d. $4000

5. Yvette wants to find her new gross pay. She should ___________
  a. multiply the answer in #4 by .06.
  b. divide the answer in #4 by .06.
  c. subtract the answer in #4 from 24,000.
  d. add the answer in #4 to 24,000.

Work Problem

On July 1, Vern Washington will receive an increase of 5.25%. Find Vern's new gross earnings so the amount can be recorded in the ledger. (Use the payroll summary at the beginning of this lesson.)

Answer: ___________________

Answers to Problem-Solving Practice Questions

DEFINE the problem

- The amount of Philip's raise
- To find an amount

PLAN the solution

- The percent of increase and Philip's current earnings
- The percent of increase is on the memo. Philip's current earnings are in the payroll ledger.
- The percent of increase and Philip's current earnings
- Multiplication

SOLVE the problem

- The percent of increase and Philip's current earnings
- $$\begin{array}{r} 308 \\ \times\ .05 \\ \hline \end{array}$$
- 15.40
- .05
- 15.40
- $$\begin{array}{r} .05 \\ 308\overline{)15.40} \\ \underline{15.40} \\ 0 \end{array}$$

CHECK the solution

- The defined purpose was accomplished because $15.40 represents an amount.
- The solution to the work problem was reasonable because $15.40 is part of $308.

Answers to Skills Practice Problems

Exercise 1

1. 0.30 or 0.3 **2.** 0.015 **3.** 0.06 **4.** .009 **5.** 0.25
6. 1.01 **7.** 0.49 **8.** 0.75 **9.** 0.33 1/3 **10.** 0.17

Exercise 2

1. 15 **2.** 38.4 **3.** 15.26 **4.** 6.63
5. 62 **6.** 102.5 **7.** 180 **8.** 280

Exercise 3

1. 25% **2.** 4% **3.** 10% **4.** 5% **5.** 5%

Exercise 4

1.	Base: $4000	Rate: 92%	Amount: ?
2.	Base: ?	Rate: 25%	Amount: 320
3.	Base: $520	Rate: ?	Amount: $41.60
4.	Base: $850	Rate: 70	Amount: ?
5.	Base: $420	Rate: ?	Amount: $8.40

Exercise 5

1. 900 **2.** 15,820 **3.** 1500 **4.** 1580 **5.** 1728

Putting It All Together

This unit includes how to express number relationships as fractions, ratios and percents. Also, it includes how to use percents to find amounts. Use the knowledge gained from unit two to solve these two work problems. Use this year's income statement below to solve the problems.

Income Statement
Neighborhood Youth and Family Services

Child Care:	
ages B - 5	$120,000
ages 6 - 12	$50,000
Rental Fees (Pool and Gym):	$20,000
Class and Recreation Fees:	$10,000
Grants:	
private	$80,000
state	$80,000
federal	$40,000
TOTAL INCOME:	$400,000

Work Problem A

Ted must prepare a written report to the community. He asks, "What part of the total income comes from rental fees?" Express the answer as a fraction and a percent.

	Fraction	**Percent**
Answers:	________	________

Work Problem B

Ted is planning the budget for next year. He expects that the rental fees will provide the same percent of income next year as they did this year. If Ted sets up a budget of $410,000, how much money must be provided by the rental fees?

Answer: ____________________

UNIT III

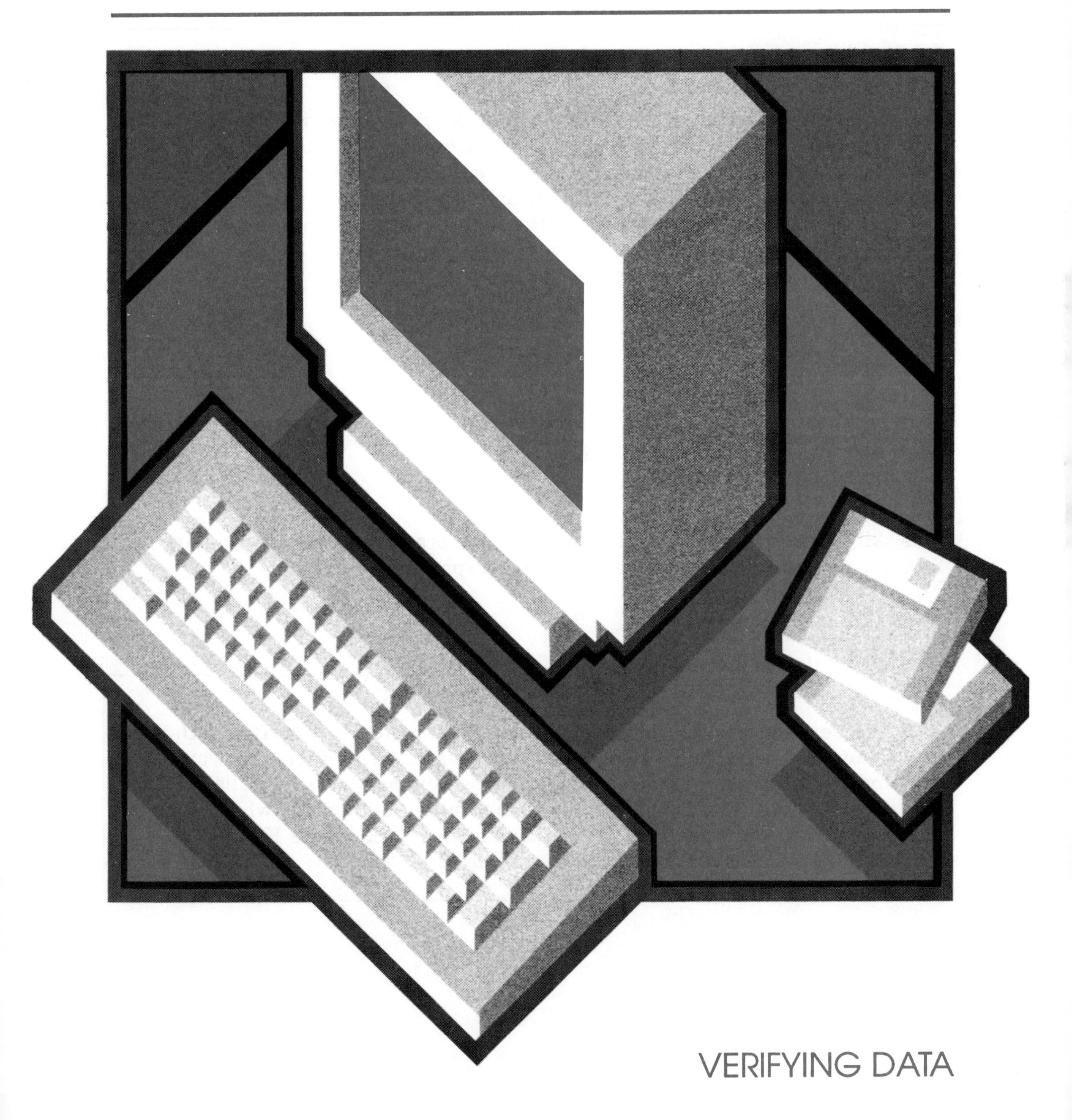

VERIFYING DATA

Office workers are frequently called upon to *verify data*, which means checking numbers and information to be sure they are accurate. Sometimes verifying requires a calculation, sometimes not. However, verifying often requires a comparison to a standard, a known amount. This unit explains how to verify that expenses are within a budgeted amount, verify the number of hours that employees work, verify data presented on graphs, and verify federal income tax deductions by using a table. When verifying data, the steps to follow are

1. Select the data.
2. Set up the calculation.
3. Do the calculation.
4. Check the accuracy of the answer.
5. Identify the standard.
6. Equate the units of measure.
7. Compare the amounts.

The skills practice exercises in this unit provide practice with comparing amounts, working with units of measure, and reading graphs and tables.

Count On Us Computers

The setting for the work problems in this unit is Count On Us Computers, a computer store which sells computers, printers, software, and other computer products. Count On Us Computers also installs and repairs computers. Employees at Count On Us Computers who often verify data are Mark Schmidt, an accounting clerk; Diane Kruger, an office clerk; and Curt Johnson, a payroll clerk.

LESSON 9

Verifying an Amount by Comparing Values

Many different types of jobs require employees to verify data. Before a company pays its bills, employees verify that the charges are valid and the amounts are accurate. Office workers verify amounts when they check the money and receipts in the petty cash box. When they compare the amount of inventory on the shelf or in the warehouse to the amount on record, employees are verifying data. Mark Schmidt, an accounting clerk at Count On Us Computers, verifies data when he confirms that a department's expenses are within budget. (And students verify data when they check their math answers against those provided at the end of a lesson.)

Mark's solution to the following work problem is shown as an example.

Work Problem

Rachel Peters, the accountant at Count On Us Computers, said to Mark, "Please verify that the January expenses of the training department were within a budget of $2000." Mark used the expense summary shown below.

Count on Us Computers

Training Department

YEARLY EXPENSE SUMMARY

MONTH	PHONE	MEALS	TRAVEL	HOTEL	ENTERTAINMENT	MISC.	MONTHLY TOTAL
JANUARY	206.80	447.90	402.20	723.30	0	46.70	
FEBRUARY	230.20	502.40	447.80	800.20	0	66.50	
MARCH	298.00	613.20	480.00	876.50	27.00	80.00	
1ST QUARTER TOTAL							

Mark added the amounts for January. Add the amounts and write the answer below.

$$
\begin{array}{r}
\$206.80 \\
447.90 \\
402.20 \\
723.30 \\
+\ \ 46.70 \\
\hline
\end{array}
$$

Did the training department stay under budget? ________

Mark's total was $1,826.90. He compared this amount to the $2000 budget amount. Because $1,826.90 is less than $2000, Mark told Rachel, "Yes, the training department stayed under budget." Students who obtained a January total different from Mark's should review Lessons 1 and 2 in Unit I. Additional assistance and practice in comparing numbers and verifying is provided in the Skills Practice section of this lesson beginning on page 202.

Doing Math to Verify Data

Verifying an amount often requires comparing the amount to a standard. The amount is the number being checked. The standard is a known amount that is used as a comparison.

Mark knows that his purpose is to compare the January expense total (the amount) with the budget of $2,000 (the standard). He knows the individual January expenses and knows that he must add them and compare the total to $2,000. He follows the solution steps below.

1. Select the relevant data. *The expenses for January*

2. Set up the calculation.

$$
\begin{array}{r}
\$\ 206.80 \\
447.90 \\
402.20 \\
723.30 \\
+\ \ 46.70 \\
\hline
\end{array}
$$

3. Do the calculation. *$1826.90*

4. Check the accuracy of the answer.

```
$   46.70
   723.30
   402.20
   447.90
+  206.80
$1826.90
```

5. Identify the standard of comparison.

$2000.00

6. Equate the units of measure.

In this example, no conversion is necessary. Both numbers are dollar amounts.

Note: Sometimes the amounts being compared are expressed in different *units*. For example, one may be a decimal and the other a fraction. Or perhaps the amount is in feet and the standard is in yards. When the units are different, either the amount or the standard must be changed so that the units are the same type.

7. Compare the amounts.

Since $1826.90 is less than $2000.00, the training department stayed within its budget.

Now an entire problem will be completed.

Work Problem

The training department budgeted $1500 for meal expenses during the first quarter. At the end of the quarter, Rachel said to Mark, "Did the training department stay within its meal budget?"

DEFINE the problem

- What is the expected outcome? *Verification that the training department stayed within its budget*
- What is the purpose? *To verify an amount*

PLAN the solution

- What data are needed? *Budgeted amount (standard) and 1st quarter meal expenses (amount)*
- Where can the data be found? *Quarterly meal expenses calculated from the monthly meal expenses given; budgeted amount given*
- What is already known? *The monthly meal expenses (1st quarter) and the budgeted amount*
- Which operations should be used? *Addition*

SOLVE the problem

- Select the relevant data. *Monthly meal expenses*
- Set up the calculation.

$$\begin{array}{r} \$447.90 \\ 502.40 \\ +\ 613.20 \\ \hline \end{array}$$

- Do the calculation. *$1,563.50*
- Check the accuracy of the answer.

$$\begin{array}{r} \$613.20 \\ 502.40 \\ +\ 447.90 \\ \hline \$1{,}563.50 \end{array}$$

- Identify the standard. *Budgeted amount, $1500*

- Equate the units of measure. *Units already the same (dollars)*

- Compare the amounts. *$1,563.50 is greater than $1,500; training department did not stay within budget.*

CHECK the solution

Make sure the solution solves the work problem.

- Was the defined purpose accomplished? *Yes, expenses were verified.*

- Is the solution to the work problem reasonable? *Yes, $1563.50 likely; it is near the budgeted amount of $1,500.*

Problem-Solving Practice

Use the DEFINE, PLAN, SOLVE, and CHECK steps to solve the work problem below.

Work Problem

Rachel asks an accounting clerk at Count On Us Computers, "Did the training department stay within its $2000 budget for February?"

DEFINE the problem

- What is the expected outcome?

- What is the purpose? ______________________________

PLAN the solution

- What data are needed? ______________________________

- Where can the data be found? ______________________________

- What is already known? ______________________________

- Which operations should be used? ______________________________

SOLVE the problem

- Select the relevant data. ______________________________

- Set up the calculation.

- Do the calculation.

- Check the accuracy of the answer.

- Identify the standard. ______________________

- Equate the units of measure. ______________________

- Compare the amounts. ______________________

____

____

CHECK the solution

Make sure the solution solves the work problem.

- Was the defined purpose accomplished? ______________________

____

____

- Is the solution to the work problem reasonable? ______________________

____

____

Answers to Problem-Solving Practice questions appear on page 208.

On Your Own

Solve the work problems below. Remember to DEFINE the problem, PLAN the solution, SOLVE the problem, and CHECK the solution. Use the expense summary found at the beginning of this lesson for Work Problems A and B.

Work Problem A

Rachel is working on a first quarter financial report. She says to Mark, "The training department budgeted $750 for phone expenses for the first quarter. Verify that the training department stayed within its budget."

Answer: ____________________

Work Problem B

Rachel is working on the monthly financial report. She says, "The training department worked with a budget of $2400 for March. Did it stay within its budget?"

Answer: ____________________

Work Problem C

Rachel is working on the monthly financial report for June. She says, "Find out if the training department stayed within its budget of $2600 this month. Also, please tell me the difference between the budgeted amount and the actual expenses." Use the expense summary shown below for Work Problems C, D, and E.

Answer: ____

Count on Us Computers

Training Department

YEARLY EXPENSE SUMMARY

MONTH	PHONE	MEALS	TRAVEL	HOTEL	ENTERTAINMENT	MISC.	MONTHLY TOTAL
JANUARY	206.80	447.90	402.20	723.30	0	46.70	
FEBRUARY	230.20	502.40	447.80	800.20	0	66.50	
MARCH	298.00	613.20	480.00	876.50	27.00	80.00	
1ST QUARTER TOTAL							
APRIL	212.20	495.90	414.00	719.00	0	50.10	
MAY	242.80	510.10	459.30	789.70	0	71.80	
JUNE	301.00	674.00	503.90	905.80	32.00	92.20	
2ND QUARTER TOTAL							

Work Problem D

The training department had budgeted $1500 for travel expenses during the second quarter. Rachel says, "Find out if the training department stayed within its travel budget. Also, tell me the difference between the budgeted amount and the actual expenses."

Answer: ______________________________

Work Problem E

During the second quarter the training department budgeted $800 for phone expenses. Rachel would like to know if it stayed within its budget. She also asks, "What is the difference between the actual and budgeted amounts? What percent of the budget amount is that difference?" (To determine that percent, called the *percent difference*, divide the difference by the budgeted amount.)

Answer: ______________________________

Skills Practice: Verifying Amounts

Office workers are often required to verify amounts listed on different documents. For example, verifying expenses may require comparing an expense statement to invoices or receipts. Verifying a bill may require comparing an invoice number on the bill to the number on the invoice. The exercises below provide practice with comparing numbers and amounts.

Exercise 1

Compare the purchase order numbers in the column A to those in column B. Circle the mistakes in column B.

	A	B
1.	95675	95675
2.	67268	67268
3.	57645	57465
4.	66585	66585
5.	48508	48585
6.	76818	76818
7.	01861	01861
8.	31121	31121
9.	22263	22262
10.	30757	30757
11.	41581	45181
12.	36623	36626
13.	91808	91808
14.	51231	51231
15.	92499	94299
16.	56675	56675
17.	66165	66165
18.	84468	84468
19.	51505	50505
20.	52242	52242
21.	90880	90880
22.	41381	41380
23.	25544	25544
24.	64405	64045
25.	86276	86276
26.	93797	93797
27.	28827	28287
28.	95675	95675
29.	60790	60790
30.	79447	79447

Answers begin on page 209.

Exercise 2

Verify that the account numbers in column A are the same as those in column B. If the numbers are the same, circle "OK" at the end of each problem. If the numbers are not the same, circle the numbers in column B that need correction.

	A	B
1.	4086	4086
	4882	4882
	8064	8064
	0485	0465
	OK	
2.	8020	8020
	2040	2040
	4808	4880
	0180	0180
	OK	
3.	3012	3012
	2113	2013
	3230	3230
	1230	1230
	OK	
4.	7982	7982
	0789	0789
	7509	7509
	7977	7977
	OK	
5.	8404	8404
	4484	4484
	7180	7184
	8417	8417
	OK	
6.	2587	2587
	8917	8917
	5188	5188
	7952	7952
	OK	

Answers begin on page 209.

Exercise 3

On checks, dollar amounts are written in words and numbers.

First State Bank
1234 Main Street
Anywhere, USA 12345

Date ____________

Pay to the
Order of ____________ Amount 17.70

$ Seventeen and 70/100 ———— dollars

____

____

1234 123 12 1234   12121 212 222 1 2312

Verify that the amounts in column B are correct by comparing them with the amounts written in column A. Circle the numbers in column B that are *not* correct.

	A	B
1.	Seventeen and 7/100	17.70
2.	Ten and 17/100	10.17
3.	Seventy and 10/100	70.10
4.	Seventeen and 70/100	70.70
5.	Seventeen and 10/100	17.10
6.	Ten and 7/100	10.07
7.	Three and 50/100	3.50
8.	Thirty-five and 3/100	35.03
9.	Five and 33/100	5.53
10.	Thirty-three and 5/100	33.05
11.	Three and 53/100	3.53
12.	Five and 30/100	5.03

Answers begin on page 209.

Exercise 4

The two lists below contain the same account numbers, written in a different order. However, a number is missing from list B. Write the missing number on the blank provided.

A	B
5012	5130
5203	5230
5302	5212
5130	5012
5023	5002
5320	5103
5103	5203
5212	5302
5230	5320
5002	______

Answers begin on page 209.

Exercise 5

The two lists below contain the same dollar amounts, written in a different order. However, an amount is missing from list B. Write the missing amount on the blank provided.

A	B
$2742.80	$2927.30
$1860.30	$1946.80
$3478.00	$2742.80
$2927.30	$3298.00
$2814.70	$1706.50
$1706.50	$3145.50
$2487.50	$3478.00
$1946.80	$2814.70
$3145.50	$1860.30
$3298.00	______

Answers begin on page 209.

Exercise 6

1. Circle the amounts greater than 3010.

998	1100	3103	2867	3009	1030

2. Circle the amounts greater than 7.99.

17	5.23	8.1	32.3	26	.1799

3. Circle the amounts less than 6.25.

.956	3.926	.789	10.4	33.8	3.83

4. Circle the amounts less than 12.31.

20.6	1.98	9.65	7.32	19	2.06

5. Circle the amounts greater than 9.1.

5.87	11	24	9.06	6.96	2.4

6. Circle the amounts equal to .05. Some amounts need to be converted to decimal format.

.050	.005	.5	5/100	5/10	50/1000

7. Circle the amounts less than 75%. Some amounts need to be converted to percent format.

69%	9/10	8.5%	101%	.8	.39

Check Yourself

Circle the letter of the correct answers.

1. Which of the following situations require an employee to verify an amount?
   a. Determining monthly totals
   b. Determining quarterly totals
   c. Determining how much money has been spent from petty cash
   d. Checking that the amount in petty cash is the same as the amount in the record book

2. Verifying data requires a comparison between ________
   a. a part and a whole.
   b. a number of parts and 100.
   c. a standard and an amount.
   d. a total and a difference.

3. A standard ________________
   a. expresses a relationship.
   b. is used to compare or to check data.
   c. is a unit of measure.
   d. is a problem solution.

Refer to the expense statement below to answer questions 4 and 5.

Count on Us Computers

Training Department

YEARLY EXPENSE SUMMARY

MONTH	PHONE	MEALS	TRAVEL	HOTEL	ENTERTAINMENT	MISC.	MONTHLY TOTAL
JANUARY	206.80	447.90	402.20	723.30	0	46.70	1826.90
FEBRUARY	230.20	502.40	447.80	800.20	0	66.50	2047.10
MARCH	298.00	613.20	480.00	876.50	27.00	80.00	2374.70
1ST QUARTER TOTAL	735.00	1563.15	1330.00	2400.00	27.00	193.20	6248.70
APRIL	212.20	495.90	414.00	719.00	0	50.10	1895.20
MAY	242.80	510.10	459.30	789.70	0	71.80	
JUNE	301.00	674.00	503.90	905.80	32.00	92.20	
2ND QUARTER TOTAL							

4. Rachel asked Mark to verify that the training department stayed within a budget of $2000 during April. What is the standard in this problem?

a. $6248.70
b. $2000.00
c. $1895.20
d. $719.00

5. What is the amount to be compared to the standard?

a. $6248.70
b. $2000.00
c. $1895.20
d. $719.00

Work Problem

The training department budgeted $2500 for hotel expenses during the second quarter. Rachel is preparing the second quarter report. She asks, "Did the training department stay within its hotel budget?" Use the information in the expense statement on page 207 to solve this work problem.

Answer: ____________________

Answers to Problem-Solving Practice Questions

DEFINE the problem

- Verification of the training department's total expenses for February
- To verify an amount

PLAN the solution

- The budgeted amount and the training department's total expenses for February
- The budgeted amount was given in the problem. The training department's total expenses can be calculated from the individual expenses for February provided on the expense statement.
- The budgeted amount and the individual expenses for February are already known.
- Addition

SOLVE the problem

- The individual expenses for February.
- $ 230.20
502.40
447.80
800.20
+ 66.50
- $2047.10
- $ 66.50
800.20
447.80
502.40
+ 230.20
$2047.10
- The standard is $2000.
- The units are already the same.
- The dollar amount, $2047.10, is greater than $2000. The training department spent more than its budget.

CHECK the solution

- The purpose was accomplished. A verification was made.
- The solution to the work problem is reasonable since $2047.10 is near the budgeted amount of $2000. Therefore, the solution is reasonable.

Answers to Skills Practice Problems

Exercise 1

There are mistakes in items 3, 5, 9, 11, 12, 15, 19, 22, 24 and 27.

Exercise 2

1. 0465 should be circled.
2. 4880 should be circled.
3. 2013 should be circled.
4. ok
5. 7184 should be circled.
6. ok

Exercise 3

There are mistakes in 1, 4, 9, 12.

Exercise 4

5023

Exercise 5

$2487.50

LESSON 10

Verifying Measures of Time, Length, and Weight

Time measurement is critical to many businesses. The most common reason to measure time is so that employees are paid properly. Because many employees are paid based on how much time they work, time computations must be very accurate. At the end of a payroll period, the number of hours worked by an employee is verified by a payroll clerk. Accounting clerks, like Mark Schmidt, sometimes verify the time an employee spent doing a repair or installation so that the company can bill a customer for the service.

Mark's solution to the following work problem is shown as an example.

Work Problem

At Count On Us Computers, each employee turns in a time sheet at the end of the week. Below is a partially completed time sheet for Tim Holt, a computer technician. On March 7, Tim spent all day repairing computers at First State Bank. He told Mark that First State Bank should be billed for $5\frac{1}{4}$ hours of labor. Mark must verify this amount before sending the bill.

WEEKLY TIME SHEET

NAME: *Tim Holt* **DEPT.** *Service*

PAY PERIOD BEGINNING: March 4, 19--

	Morning		Afternoon		Total	
DATE	**Hours**	**Minutes**	**Hours**	**Minutes**	**Hours**	**Minutes**
3/4	3	—	4	—	7	—
3/5	3	30	3	—	6	30
3/6	2	30	4	—	6	30
3/7	2	45	2	30		
3/8	2	30	3	30		
				Total for Payroll Period		

First, Mark added Tim's hours. Add and write the answer below.

2 hours 45 minutes
+ 2 hours 30 minutes

Mark determined that Tim worked 5 hours and 15 minutes. To compare that amount with the standard of 5¼ hours, Mark needed the same kind of units of measure; he needed to express 15 minutes as a part of an hour. He set up the fraction, $\frac{15}{60}$, and reduced it to lowest terms, ¼. Mark could see that Tim was correct, and he billed First State Bank for 5¼ hours of Tim's labor.

Students who did not obtain an answer of 5¼ hours or 5 hours and 15 minutes may need more practice adding and converting units of time. The Skills Practice section beginning on page 220 provides assistance and practice calculating with units of time.

Doing Math to Verify Data

Mark needed to know how many hours Tim worked before he could verify that 5¼ was correct. In order to find the total number of hours that Tim worked on March 7, Mark planned to add the time worked in the morning and the afternoon. He also planned to convert that total to hours and fractions of an hour (from hours and minutes).

1. **Select the relevant data.** — *The morning and afternoon time Tim worked on March 7*

2. **Set up the calculation.**

   2 hours 45 minutes
   + 2 hours 30 minutes

3. **Do the calculation.**

   2 hours 45 minutes
   + 2 hours 30 minutes
   4 hours 75 minutes

   Convert 75 minutes to one hour and fifteen minutes.

   4 hours
   + 1 hour 15 minutes
   5 hours 15 minutes

4. **Check the accuracy of the answer.**

   5 hours 15 minutes
   − 2 hours 30 minutes
   2 hours 45 minutes

5. **Identify the standard.** — *Standard is 5¼ hours.*

6. **Equate the units of measure.** — *The standard expressed in hours. Tim's time, expressed in hours and minutes, must be converted:*

   5 hours 15 minutes = $5\frac{15}{60}$ hours = 5¼ hours

7. Compare the amounts.

Tim's time is equal to the standard, 5¼. Tim's estimate was correct. The bank must be billed for 5¼ hours.

Now watch Mark work through an entire problem.

Work Problem

Tim spent all of March 8 repairing the computers at Ace Accounting. Tim told Mark, "Send Ace Accounting a bill for 6 hours of service." Mark must verify Tim's estimate.

DEFINE the problem

- What is the expected outcome? *Verification of the number of hours Tim worked at Ace Accounting*

- What is the purpose? *To verify an amount*

PLAN the solution

- What data are needed? *The number of hours Tim wourked at Ace Accounting and Tim's estimate*

- Where can the data be found? *Number of hours can be calculated from Tim's time sheet. Tim has already given the estimate.*

- **What is already known?** *Tim's morning and afternoon hours and Tim's estimate*

- **Which operations should be used?** *Addition*

SOLVE the problem

- **Select the relevant data.** *Tim's morning and afternoon hours*

- **Set up the calculation.**

  2 hours 30 minutes
  + 3 hours 30 minutes

- **Do the calculation.**

  5 hours 60 minutes
  + 1 hour
  6 hours

- **Check the accuracy of the answer.**

  5 hours 60 minutes
  − 2 hours 30 minutes
  3 hours 30 minutes

- **Identify the standard.** *The standard is 6 hours.*

- **Equate the units of measure.** *The units are already the same.*

- **Compare the amounts.** *6 hours = 6 hours, so Tim's estimate is correct.*

CHECK the solution

Make sure the solution solves the work problem:

- Was the defined purpose accomplished?

  Yes, Tim's estimate was verified.

- Is the solution to the work problem reasonable?

  Yes, because it is same as Tim's estimate and is close to usual work day (8 hours)

Problem-Solving Practice

Use the DEFINE, PLAN, SOLVE, and CHECK steps to solve the following work problem.

Work Problem

Tim is a part-time employee. To qualify for health benefits, he must work at least 30 hours each week. The payroll clerk must verify that Tim worked 30 hours during the week of March 4. Solve this work problem for the payroll clerk.

WEEKLY TIME SHEET

NAME: Tim Holt **DEPT.** Service

PAY PERIOD BEGINNING: March 4, 19--

	Morning		Afternoon		Total	
DATE	Hours	Minutes	Hours	Minutes	Hours	Minutes
3/4	3	—	4	—	7	—
3/5	3	30	3	—	6	30
3/6	2	30	4	—	6	30
3/7	2	45	2	30	5	15
3/8	2	30	3	30	6	—
				Total for Payroll Period		

DEFINE the problem

- What is the expected outcome? ______________________
- What is the purpose? ______________________

PLAN the solution

- What data are needed? ______________________

- Where can the data be found? ______________________

____

____

- What is already known? ______________________

____

____

- Which operations should be used? ______________________

SOLVE the problem

- Select the relevant data. ______________________

____

____

- Set up the calculation.

- Do the calculation.

- Check the accuracy of the answer.

- Identify the standard. ______________________

- **Equate the units of measure.** ____________________

- **Compare the amounts.** ____________________

____

____

CHECK the solution

- **Was the defined purpose accomplished?** ____________________

____

____

- **Is the solution to the work problem reasonable?** ____________________

____

____

Answers to Problem-Solving Practice questions appear on page 230.

On Your Own

Solve the work problems below. Remember to DEFINE the problem, PLAN the solution, SOLVE the problem, and CHECK the solution.

Work Problem A

Sharon spent all of March 14 installing computers at a law firm. Sharon said, "The law firm should be billed for $5\frac{3}{4}$ hours of service." Verify Sharon's estimate.

Answer: ____________________

WEEKLY TIME SHEET

NAME: Sharon Goldman DEPT. Service

PAY PERIOD BEGINNING: March 11, 19--

	Morning		Afternoon		Total	
DATE	Hours	Minutes	Hours	Minutes	Hours	Minutes
3/11	3	30	2	30	6	—
3/12	2	30	2	30	5	—
3/13	3	30	3	15	6	45
3/14	3	30	2	15		
3/15	3	15	3	15		
				Total for Payroll Period		

Work Problem B

Sharon spent all of March 15 installing the computers at an advertising agency. At the end of the day, Sharon said, "The advertising agency should be billed for 6½ hours of service." Verify Sharon's estimate.

Answer: ____________________

Work Problem C

Sharon is a part-time employee. To earn vacation time, she must work at least 20 hours each week. Verify that Sharon worked 20 hours during the week of March 11.

Answer: ____________________

Work problems D and E require calculations with units of *length* rather than units of time. The procedure is the same, however. Remember that 3 feet = 1 yard.

Work Problem D

When Sharon was at the law firm, she needed 20 feet of printer cable for one office and 40 feet of printer cable for another office. She had 25 yards of printer cable in her truck. Did Sharon have enough printer cable to finish the job?

Answer: ____________________

Work Problem E

At the advertising agency, Sharon needed 45 feet of computer cable for one floor and 60 feet of computer cable for another floor. She had 40 yards of computer cable in her truck. Did Sharon bring enough computer cable with her?

Answer: ____________________

Skills Practice: Units of Measure

CONVERTING UNITS OF MEASURE

The table below shows equivalent units of time, length, and weight. Smaller units appear on the left of the equal sign; larger units appear on the right of the equal sign.

TIME, LENGTH, AND WEIGHT

TIME

60 seconds	=	1 minute
60 minutes	=	1 hour
24 hours	=	1 day
7 days	=	1 week
365 days	=	1 year (366 days are in a leap year)
52 weeks	=	1 year
12 months	=	1 year

LENGTH

12 inches	=	1 foot
36 inches	=	1 yard
3 feet	=	1 yard
5280 feet	=	1 mile
1760 yards	=	1 mile

WEIGHT

16 ounces	=	1 pound
2000 pounds	=	1 ton

The number of smaller units in one larger unit is sometimes called a *conversion factor*. For example, when converting feet to yards, the conversion factor is *3* because there are 3 feet in 1 yard.

Exercise 1

Write the conversion factors on the lines provided.

1. inches to yards ________ inches in one yard

2. pounds to tons ________ pounds in one ton

3. hours to days ________ hours in one day

4. years to weeks ________ weeks in one year

5. feet to miles _________ feet in one mile

6. months to years _________ months in one year

7. seconds to minutes _________ seconds in one minute

8. ounces to pounds _________ ounces in one pound

Answers begin on page 231.

There are two rules to remember when converting from one unit to another:

1. To convert a smaller unit to a larger unit, divide the number of smaller units by the conversion factor. (For example, to convert 24 inches to feet, divide the number of inches by 12. The result is two feet.)

2. To convert a larger unit to a smaller unit, multiply the number of larger units by the conversion factor. (For example, to convert two feet to inches, multiply the number of feet by 12. The result is 24 inches.)

Another example using rule 1 to convert days to weeks is shown below.

Convert 38 days to weeks.

Step 1. Divide the smaller unit (38 days) by the conversion factor (7, because there are 7 days in 1 week).

$$\begin{array}{r} 5 \\ 7\overline{)38} \\ \underline{35} \\ 3 \end{array}$$

The answer is 5 weeks and 3 days. Note: The remainder, if any, is not usually carried forward to produce a decimal value for the quotient.

Another example using rule 2 to convert days to hours is shown below.

Convert 3 days to hours.

Step 1. Multiply the larger unit (3 days) by the conversion factor (24, because there are 24 hours in 1 day).

$3 \times 24 = 72$

There are 72 hours in 3 days.

Exercise 2

Make the following conversions.

1. Change 8 minutes to seconds.

2. Change 5 hours to minutes.

3. Change 8.5 days to hours.

4. Change 63 days to weeks.

5. Change 48 months to years.

6. Change 55 inches to feet.

7. Change 11 feet to yards.

8. Change 21,120 feet to miles.

9. Change 72 inches to yards.

10. Change 32 ounces to pounds.

11. Change 3.5 pounds to ounces.

12. Change 8,000 pounds to tons.

Answers begin on page 232.

ADDING UNITS OF MEASURE

Only the same kind of units can be added. For example, yards must be added to yards, and feet must be added to feet. The total is then converted to as many of the large units as possible.

The addition of 6 yards and 2 feet to 5 yards and 2 feet is shown below.

Step 1: Add units of the same kind.

```
   6 yards 2 feet
+  5 yards 2 feet
  11 yards 4 feet
```

Step 2. If possible, convert the smaller units in the answer to larger units.

In the total found in step 1, 4 feet can be converted to 1 yard and 1 foot.

Step 3. Add the result of step 2 to the larger units.

```
   6 yards 2 feet
+  5 yards 2 feet
  11 yards
+  1 yard  1 foot
  12 yards 1 foot
```

Exercise 3

Add the following.

1.
```
  5 miles 3,000 feet
+2 miles 4,920 feet
```

2.
```
  7 feet 10 inches
+2 feet 2 inches
```

3.
```
  3 weeks 6 days
+5 weeks 5 days
```

4.
```
  5 pounds  9 ounces
+7 pounds 11 ounces
```

5.
```
   12 pounds 11 ounces
+   5 pounds  7 ounces
```

6.
```
   5 hours 15 minutes
+4 hours 45 minutes
```

Answers begin on page 232.

SUBTRACTING UNITS OF MEASURE

Only the same kind of units can be subtracted. Sometimes it is necessary to borrow from the larger unit.

Subtraction of 3 hours and 50 minutes from 8 hours and 15 minutes is shown below.

Step 1. Borrow from the larger unit, if necessary. In this example, 50 minutes cannot be subtracted from 15 minutes, so 1 hour is "borrowed" from the 8, leaving 7 hours. The one hour is then converted to 60 minutes. The 60 minutes and 15 minutes are then added, resulting in 75 minutes.

```
   8 hours 15 minutes    becomes      7 hours 75 minutes
– 3 hours 50 minutes                – 3 hours 50 minutes
```

Step 2. Subtract.

```
   7 hours 75 minutes
– 3 hours 50 minutes
   4 hours 25 minutes
```

Exercise 4

Subtract the following.

1. 7 feet 9 inches
− 2 feet 4 inches

2. 8 feet 3 inches
− 2 feet 5 inches

3. 12 pounds 7 ounces
− 7 pounds 2 ounces

4. 17 pounds 3 ounces
− 8 pounds 8 ounces

5. 5 weeks 5 days
− 3 weeks 6 days

6. 5 yards 1 foot
− 2 yards 2 feet

Answers begin on page 232.

Exercise 5

1. Change 9,000 pounds to tons.

2. Change 2.5 tons to pounds.

3. Change 5 yards to feet.

4. Change 38 inches to feet.

5. Change 6.5 feet to inches

6. Change 6 years to months.

7. Change 23 weeks to days.

8. Change 7 days to hours.

9. 40 weeks 6 days
+30 weeks 5 days

10. 1,900 pounds
+1,540 pounds

11. 10 feet
− 6 feet 3 inches

12. 9 years 5 months
− 5 years 8 months

Check Yourself

Circle the letter of the correct response.

1. On Monday, Sharon worked for 6 hours and 30 minutes. On Tuesday, Sharon worked for 5 hours and 45 minutes. How much longer did she work on Monday than on Tuesday?

a. 15 minutes
b. 45 minutes
c. 1 hour, 15 minutes
d. 1 hour, 45 minutes

2. Sharon said, "I worked for 12¾ hours on Monday and Tuesday." Is Sharon correct? (Refer to problem 1 above.)

a. No
b. Yes

3. To convert from larger to smaller units, __________

a. Add.
b. Subtract.
c. Multiply.
d. Divide.

4. To convert from smaller to larger units, __________

a. Add.
b. Subtract.
c. Multiply.
d. Divide.

5. Computer cable 48 yards long equals __________

a. 4 feet.
b. 16 feet.
c. 144 feet.
d. 576 feet.

WORK Problem

Todd Fisher works as a sales assistant for Count On Us Computers. Because Todd is a minor, he is allowed to work no more than 15 hours each week. Verify that Todd did not work more than 15 hours.

Answer: ____________________

WEEKLY TIME SHEET

NAME: *Todd Fisher* **DEPT.** *Sales*

PAY PERIOD BEGINNING: March 18, 19--

	Morning		Afternoon		Total	
DATE	**Hours**	**Minutes**	**Hours**	**Minutes**	**Hours**	**Minutes**
3/18			3	—	3	—
3/19			3	15	3	15
3/20			3	15	3	15
3/21			2	30	2	30
3/22			2	30	2	30
				Total for Payroll Period		

Answers to Problem-Solving Practice Questions

DEFINE the problem

- A verification of Tim's weekly hours
- To verify an amount

PLAN the solution

- The number of hours needed to qualify and Tim's weekly hours
- The number of hours needed to qualify can be found in the problem. Tim's weekly hours can be calculated from the daily totals on his time sheet.
- The minimum number of required hours and the number of hours Tim worked each day
- Addition

SOLVE the problem

- The number of hours Tim worked each day
- 7 hours
  6 hours 30 minutes
  6 hours 30 minutes
  5 hours 15 minutes
  \+ 6 hours
- 30 hours 75 minutes

  75 minutes = 1 hour 15 minutes

  30 hours
  \+ 1 hour 15 minutes
  31 hours 15 minutes
- 6 hours
  5 hours 15 minutes
  6 hours 30 minutes
  6 hours 30 minutes
  \+ 7 hours
  30 hours 75 minutes

  75 minutes = 1 hour 15 minutes

  30 hours
  \+ 1 hour 15 minutes
  31 hours 15 minutes
- The standard is 30 hours.
- The units are already equal.
- Since 31 hours 15 minutes is greater than 30 hours, Tim worked the required number of hours.

CHECK the solution

- The defined purpose was accomplished. Tim's hours were verified.
- The result, 31 hours 15 minutes, is a likely amount for Tim's hours because the expectation is that Tim will work at least 30 hours if he wants health benefits. The solution is reasonable.

Answers to Skills Practice Problems

Exercise 1

1. 36 **2.** 2000 **3.** 24 **4.** 52
5. 5280 **6.** 12 **7.** 60 **8.** 16

Exercise 2

1. 480 seconds
2. 300 minutes
3. 204 hours
4. 9 weeks
5. 4 years
6. 4 feet, 7 inches
7. 3 yards, 2 feet
8. 4 miles
9. 2 yards
10. 2 pounds
11. 56 ounces
12. 4 tons

Exercise 3

1. 8 miles, 2640 feet
2. 10 feet
3. 9 weeks, 4 days
4. 13 pounds, 4 ounces
5. 18 pounds, 2 ounces
6. 10 hours

Exercise 4

1. 5 feet, 5 inches
2. 5 feet, 10 inches
3. 5 pounds, 5 ounces
4. 8 pounds, 11 ounces
5. 1 week, 6 days
6. 2 yards, 2 feet

LESSON 11

Verifying Data Presented on Graphs

Graphs provide a quick, clear picture of data and data relationships. Graphs can be used to compare many amounts, such as sales, profit, or expenses. Sometimes the data on graphs must be verified.

Diane Kruger's solution to the following work problem is shown as an example.

Work Problem

Rachel Peters gives monthly expense reports to Don Lai, the president of Count On Us Computers. Often Rachel shows Don the information on a graph, like the one below. Rachel asked Diane Kruger, an office clerk, to verify the graph using the amounts on the expense summary on the next page.

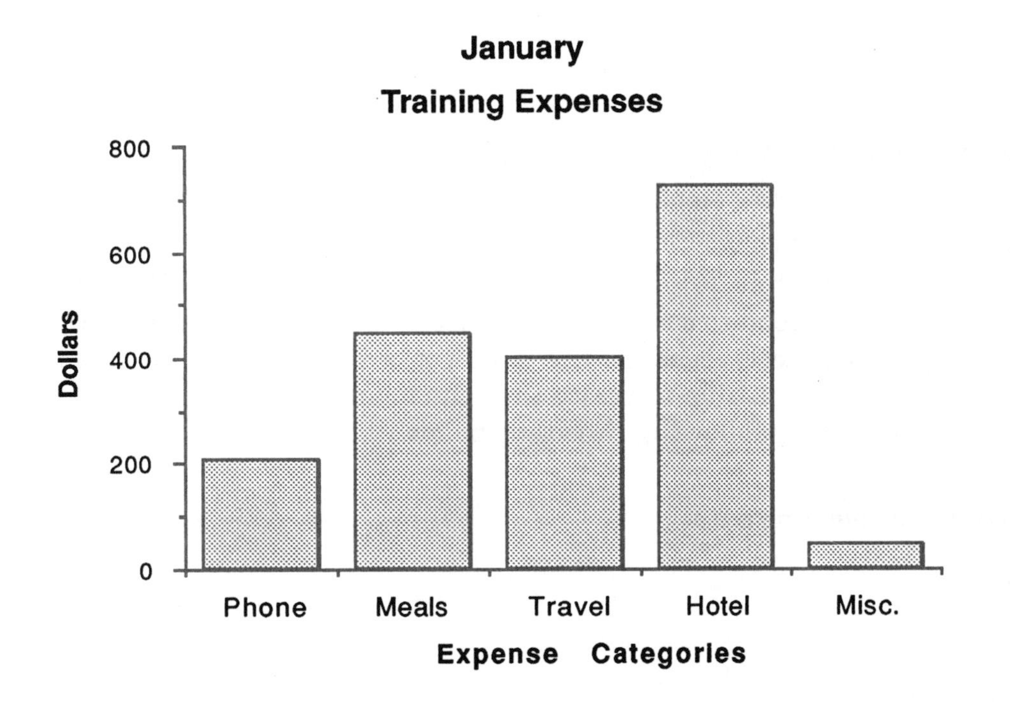

Count on Us Computers

Training Department

YEARLY EXPENSE SUMMARY

MONTH	PHONE	MEALS	TRAVEL	HOTEL	MISC.	MONTHLY TOTAL
JANUARY	206.80	447.90	402.20	723.30	46.70	
FEBRUARY	230.20	502.40	447.80	800.20	66.50	
MARCH	298.00	613.20	480.00	876.50	80.00	
1ST QUARTER TOTAL						

Diane compared the graph to the expenses for January. Do this comparison. Does the graph accurately depict the expenses?

Diane concluded that the graph was accurate. Students who concluded that the graph was not accurate, or who were not sure about its accuracy, may need more practice reading graphs. The Skills Practice section that begins on page 248 provides assistance and practice reading graphs.

Doing Math to Verify Data

Diane knew she had to verify data on the graph. She considered the standard to be the data on the expense summary. When Diane planned her solution, she realized that she would not need to calculate. She also realized that she would not be able to verify the graph exactly, because the graph is a picture without precise numbers. However, she would be able to verify that the graph looked reasonably correct. She solved the problem by following the steps below.

1. Select the relevant data. *In this problem, Diane selected the expenses shown on the graph as the relevant data.*

2. Set up the calculation. *No calculation is necessary.*

3. **Do the calculation.** — *No calculation is necessary.*

4. **Check the accuracy of the answer.** — *Does not apply.*

5. **Identify the standard.** — *The standards are the amounts listed on the expense summary for January.*

6. **Equate the units of measure.** — *The units are already the same.*

7. **Compare the amounts.** — *Diane compared each column of the bar graph to the number it represented on the expense summary. She used the numbered marks on the left side of the graph to judge whether the columns were the right height. She also compared the column heights to each other.*

She expected the "hotel" column to be the tallest because hotel expense was the greatest amount on the expense summary. She expected "meals" and "travel" columns to be about the same height, with the column for meals slightly taller. She expected the "phone" column to be about half the size of the "travel" column. Finally, she expected the "miscellaneous" column to be the shortest because this was the smallest amount on the summary. The graph looked reasonable.

Now an entire problem will be completed.

Work Problem

Rachel was preparing a report of first quarter expenses. She said to Diane, "Does this graph look like a reasonable picture of our phone expenses for the first quarter?"

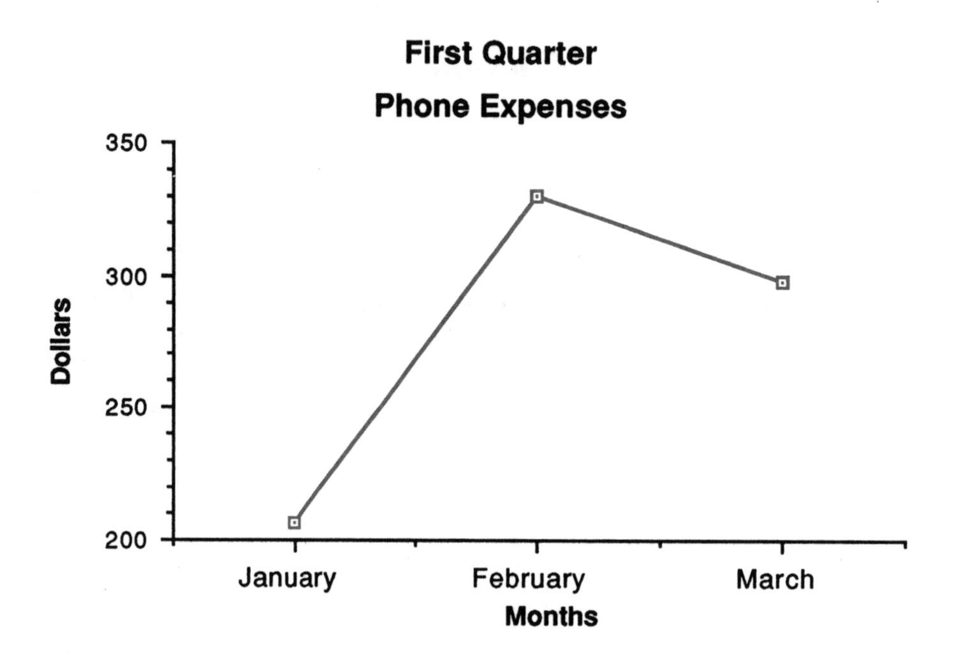

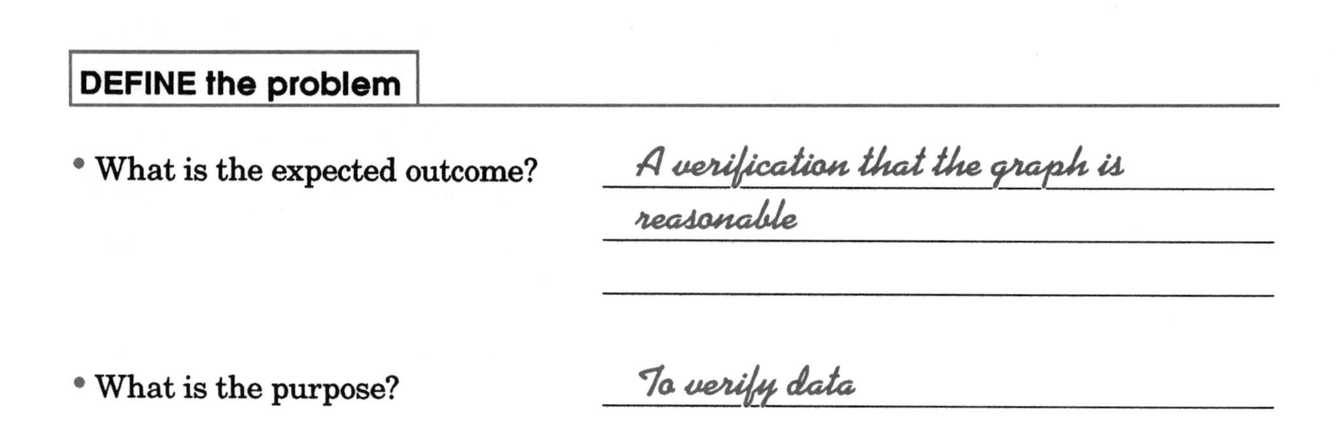

DEFINE the problem

- What is the expected outcome? *A verification that the graph is reasonable*

- What is the purpose? *To verify data*

PLAN the solution

- What data are needed? *Phone expenses for January, February and March*
- Where can the data be found? *On the expenses summary and the graph*
- What is already known? *Phone expenses for January, February and March from the expense summary and the graph*
- Which operations should be used? *No calculation necessary*

SOLVE the problem

- Select the relvelant data? *The expenses shown on the graph*
- Set up the calculation. *No calculation is needed*
- Do the calculation. *No calculation is needed*
- Check the accuracy of the answer. *Does not apply*

- Identify the standard.

  The phone expenses for January, February and March: $206.80, $230.20, $298.00.

- Equate the units of measure.

  Units are already the same

- Compare the amounts.

  January should be a little over $200. February should be almost 1/3 of the way between $200 and $300. March should be very close to $300. There is a mistake in the February expense. On the graph it appears to be over $300. The graph does not look reasonably correct.

CHECK the solution

Make sure the solution solves the work problem.

- Was the defined purpose accomplished?

  Yes, the graph was verified and a mistake was found.

- Is the solution to the work problem reasonable?

  Conclusion that the graph contains an error is reasonable.

Problem-Solving Practice

Use the DEFINE, PLAN, SOLVE, and CHECK steps to solve the next problem.

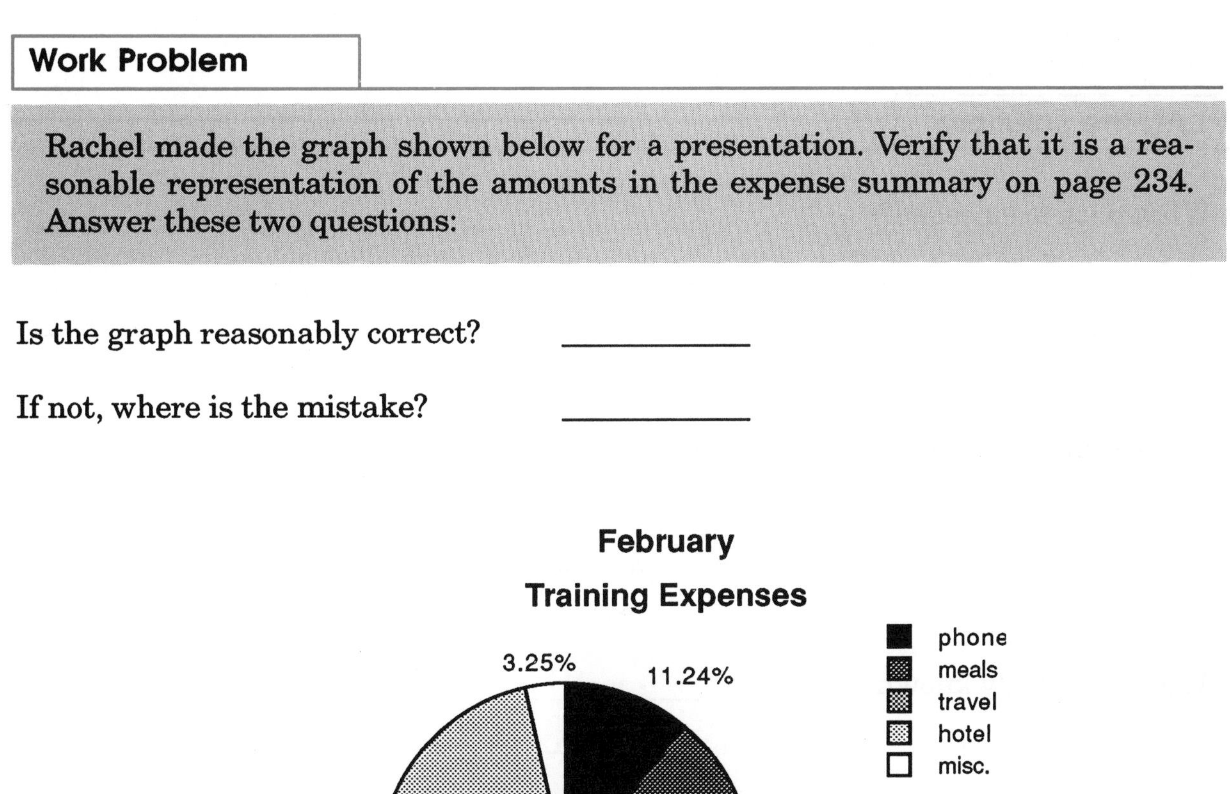

Work Problem

Rachel made the graph shown below for a presentation. Verify that it is a reasonable representation of the amounts in the expense summary on page 234. Answer these two questions:

Is the graph reasonably correct? ____________

If not, where is the mistake? ____________

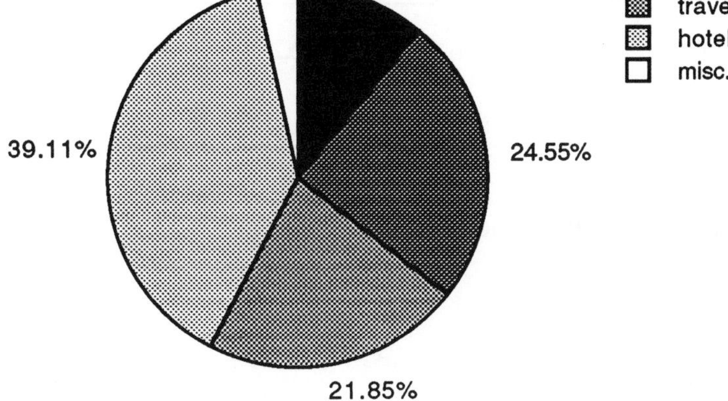

DEFINE the problem

- What is the expected outcome? ____________________________________
____
____

- What is the purpose? ______________________

PLAN the solution

- What data are needed? ______________________

- Where can the data be found? ______________________

- What is already known? ______________________

- Which operations should be used? ______________________

SOLVE the problem

- Select the relevant data. ______________________

- Set up the calculation.

- Do the calculation.

- Check the accuracy of the answer.

- Identify the standard. ______________________

- Equate the units of measure. ______________________

- Compare the amounts. ______________________

CHECK the solution

Make sure the solution solves the work problem.

- Was the defined purpose accomplished? ______________________

- Is the solution to the work problem reasonable? ______________________

Answers to the Problem-Solving Practice questions appear on page 256.

On Your Own

Solve the following work problems. Remember to DEFINE the problem, PLAN the solution, SOLVE the problem, and CHECK the solution. Use this expense summary to verify the graphs.

Count on Us Computers

Training Department

YEARLY EXPENSE SUMMARY

MONTH	PHONE	MEALS	TRAVEL	HOTEL	MISC.	MONTHLY TOTAL
JANUARY	206.80	447.90	402.20	723.30	46.70	1826.90
FEBRUARY	230.20	502.40	447.80	800.20	66.50	2047.10
MARCH	298.00	613.20	480.00	876.50	80.00	2347.70
1ST QUARTER TOTAL	735.00	1563.15	1330.00	2400.00	193.20	6221.70
APRIL	212.20	495.90	414.00	719.00	50.10	1895.20
MAY	242.80	510.10	459.30	789.70	71.80	2073.70
JUNE	301.00	674.00	503.90	905.80	92.20	2476.90
2ND QUARTER TOTAL	760.00	1680.00	1377.20	2414.50	214.10	6477.80

Work Problem A

Rachel prepared the graph shown below for the president of the company. Verify that the graph is a reasonable representation of the April expenses.

Is the graph reasonably correct? ______________________________

If not, where is the mistake? ______________________________

____

____

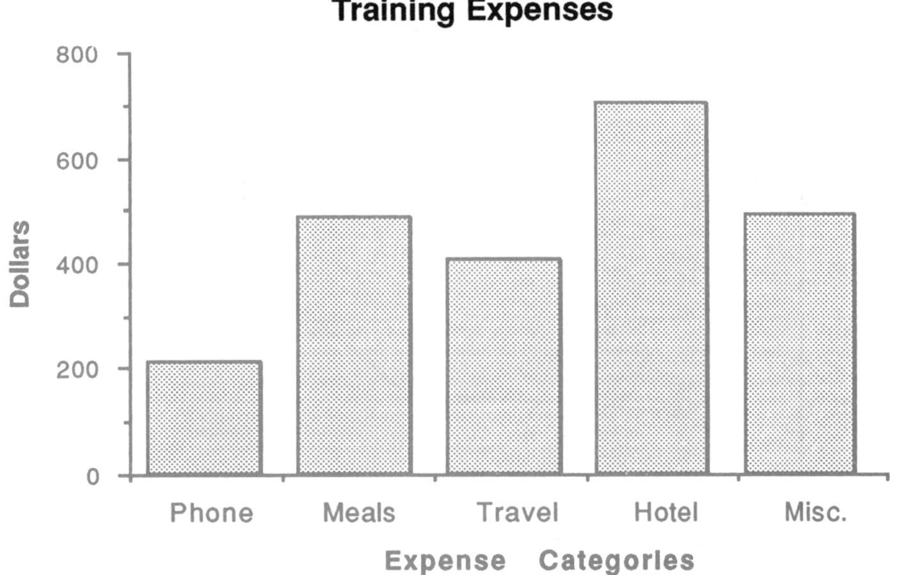

Work Problem B

Rachel prepared the graph shown below as part of a second quarter report. Verify that the graph is a reasonable representation of the meal expenses for the second quarter.

Is the graph reasonably correct? ____________________

If not, where is the mistake? ____________________

____

____

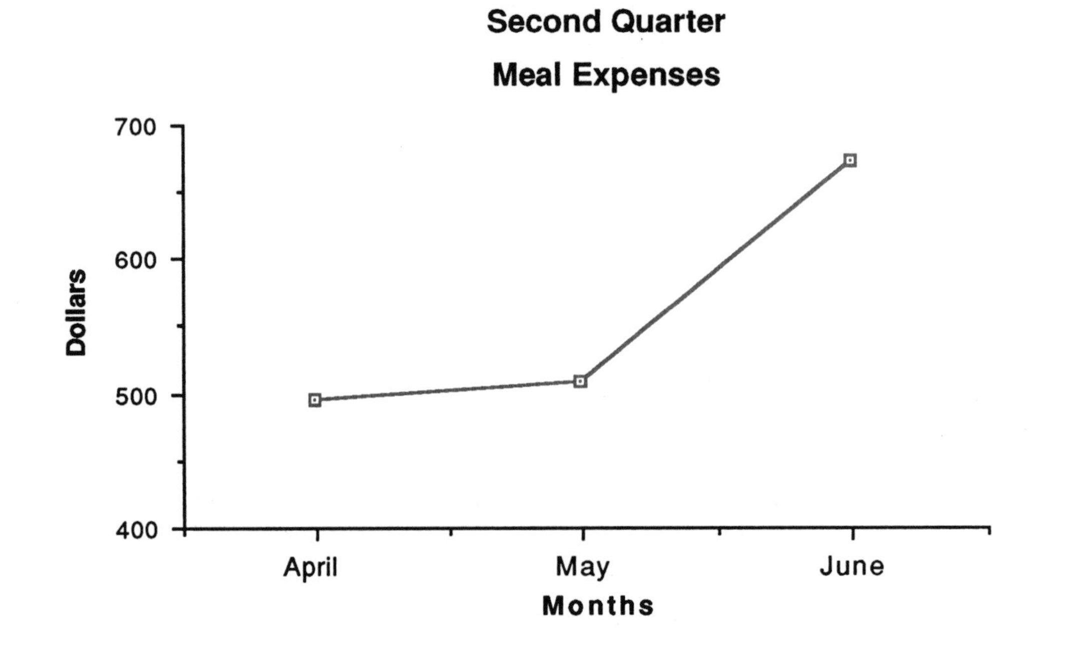

Work Problem C

One of the trainers created the graph shown below to show the training department's expenses in May. Verify that the graph is a reasonable representation of the training department's expenses during May.

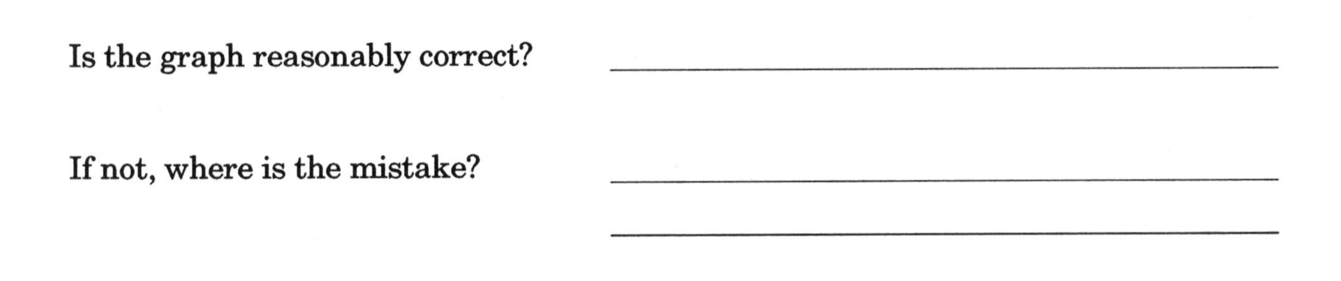

Is the graph reasonably correct? ______________________

If not, where is the mistake? ______________________

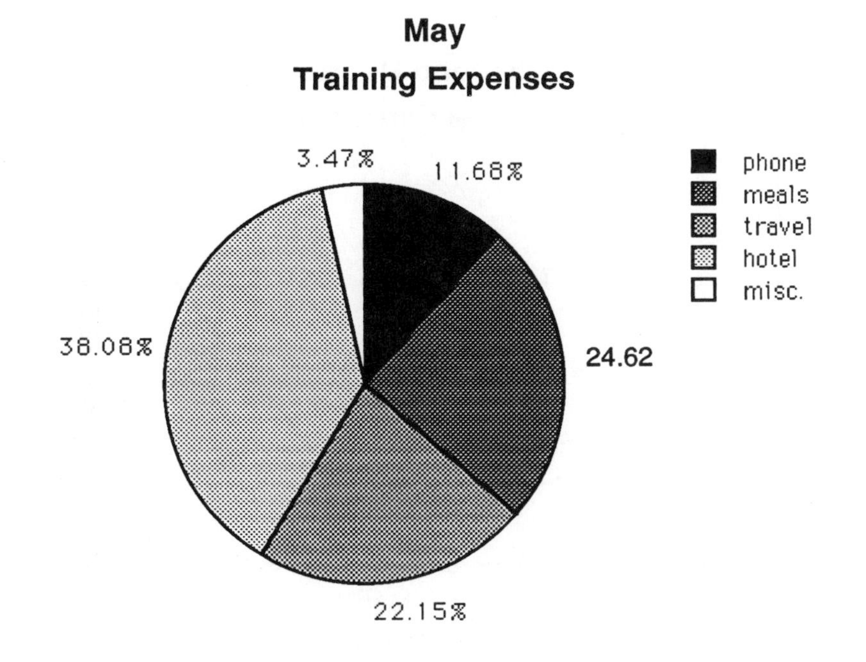

Work Problem D

Another trainer created the graph shown below to show the training department's hotel expenses. Verify that the graph is a reasonable representation of hotel expenses during the first quarter.

Is the graph reasonably correct? ______________________

If not, where is the mistake? ______________________

____

____

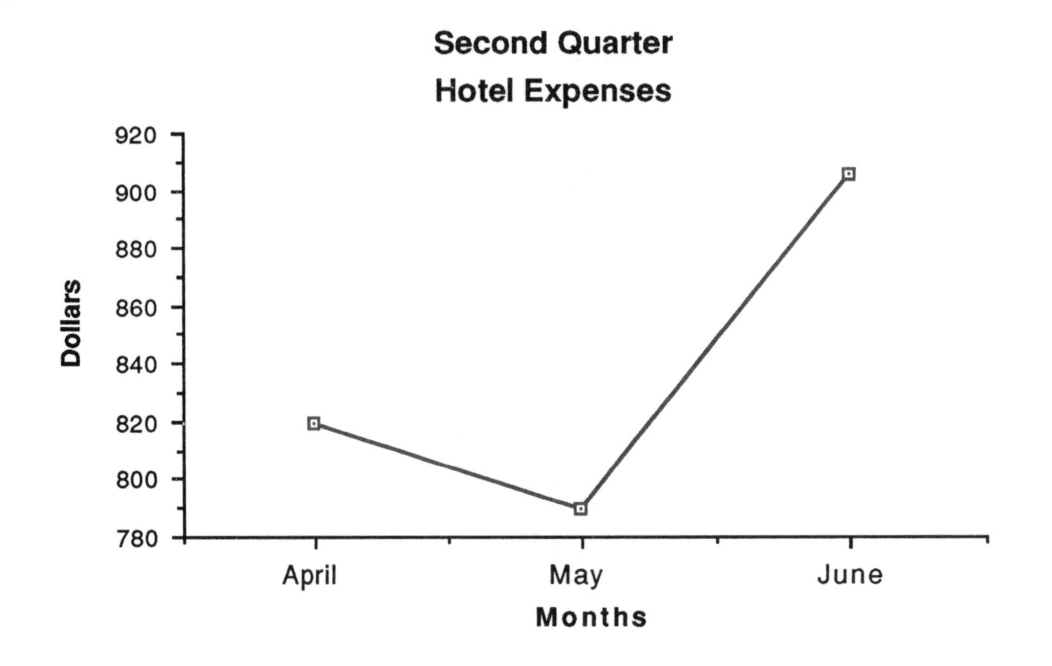

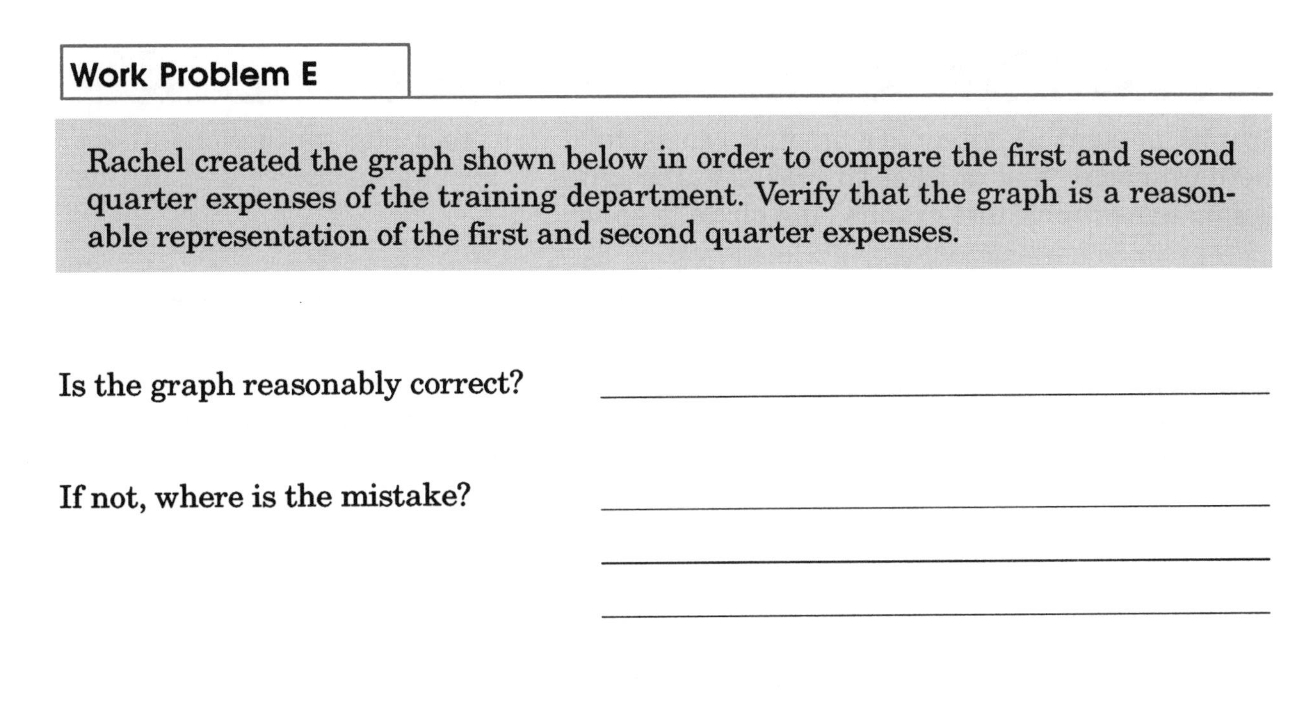

Work Problem E

Rachel created the graph shown below in order to compare the first and second quarter expenses of the training department. Verify that the graph is a reasonable representation of the first and second quarter expenses.

Is the graph reasonably correct? ____________________

If not, where is the mistake? ____________________

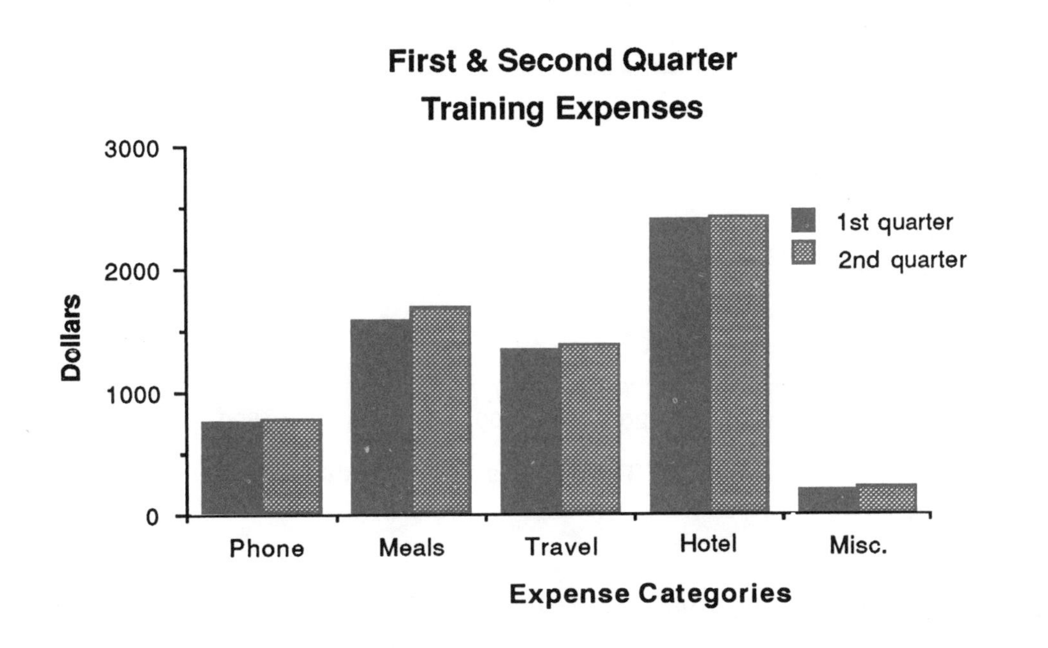

Skills Practice: Reading Graphs

Graphs present approximate amounts in picture form to make conclusions about the data easier and quicker to reach. Three kinds of graphs are presented in this lesson: bar graphs, line graphs, and circle graphs.

BAR GRAPHS

A *bar graph* uses bars to indicate amounts. Bar graphs have two *axes*, a *horizontal axis* (the line that goes across) and a *vertical axis* (the line that goes up and down). One of the axes acts as a base for the bars. The bars on the graph shown below run across, but they could run up and down, as shown earlier in this lesson.

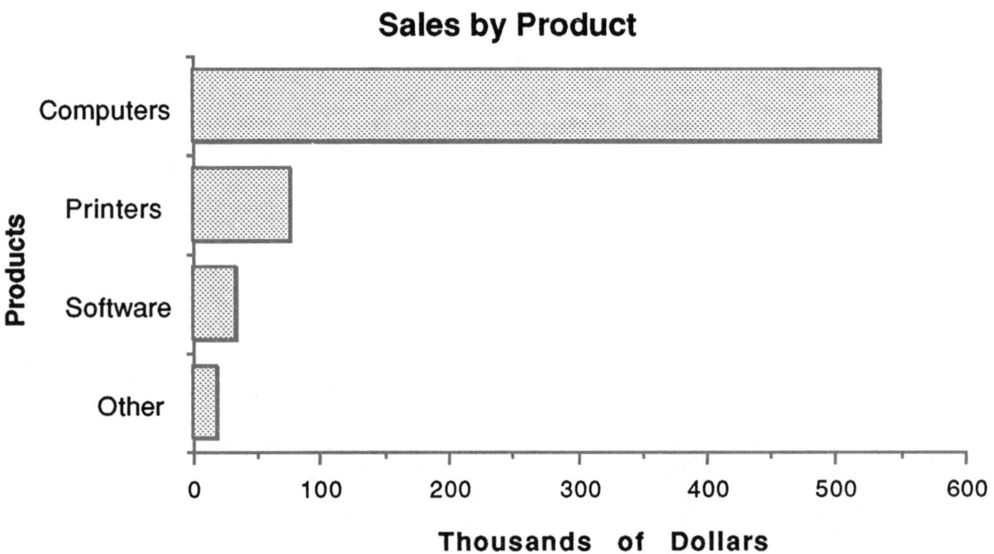

When reading a graph, always read the title first. The title should tell what the graph is about. This graph shows the sales of four product groups during January. The axes indicate what kind of information is being reported. On the bar graph above, the vertical axis shows the product groups and the horizontal axis shows a scale of dollar amounts.

Bar graphs are especially useful for making comparisons. This graph compares computers, printers, software, and other products. The product with the highest or lowest sales (dollar amount) can be identified at a glance.

To estimate "how many dollars worth" of computers were sold, compare the end of the computer bar to the dollar scale on the horizontal axis. Notice that the dollar scale is in *thousands* of dollars. The sale of computers during January were slightly less than $550,000.

Exercise 1

Study the bar graph below and answer the questions. Count On Us Computers developed the bar graph below that shows sales information for the last six months of the year.

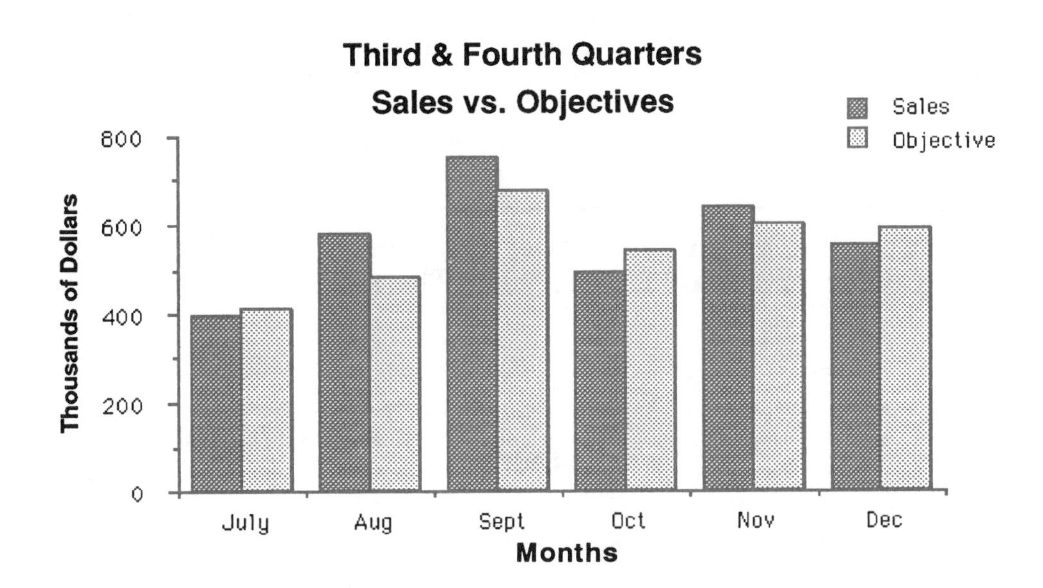

1. In which month were sales the highest? ______________________

2. In which month were sales the lowest? ______________________

3. In which months did the company meet or exceed the sales objective? ______________________ ______________________

4. Which month shows the greatest difference between the sales objective and the actual sales? ______________________

5. What type of information is presented on the horizontal axis? ______________________

6. What type of information is presented on the vertical axis? ____________________

7. What was the approximate sales amount for July? $ ____________________

Answers begin on page 257.

LINE GRAPHS

A *line graph* usually shows changes in amounts over a period of time. The lines are formed by connecting the points which represent the amounts. Like bar graphs, line graphs have horizontal and vertical axes. The horizontal axis usually indicates a period of time.

Exercise 2

Study the following line graph and answer the questions. This line graph shows Count On Us Computer's sales for one year. The sales are measured in dollar amounts.

1. In which month were sales the highest? ____________________

2. In which month were sales the lowest? ____________________

3. Between which two months did sales increase the most? ____________________________

____

4. Between which two months did sales decrease the most? ____________________________

____

5. In general, how did the amounts change from April to July? In other words, what kind of trend is noticeable? ____________________________

____

____

6. Approximately, what were the sales in September? $ ____________________________

Answers begin on page 257.

CIRCLE GRAPHS

A *circle graph*, or pie graph, shows how an amount is divided. The circle represents 100% of the amount, and each section represents a part of the amount.

Exercise 3

Study the pie graph below and answer the questions that follow it. In this circle graph, the sales total for the first quarter is divided according to product groups. Each product group represents a section of the circle.

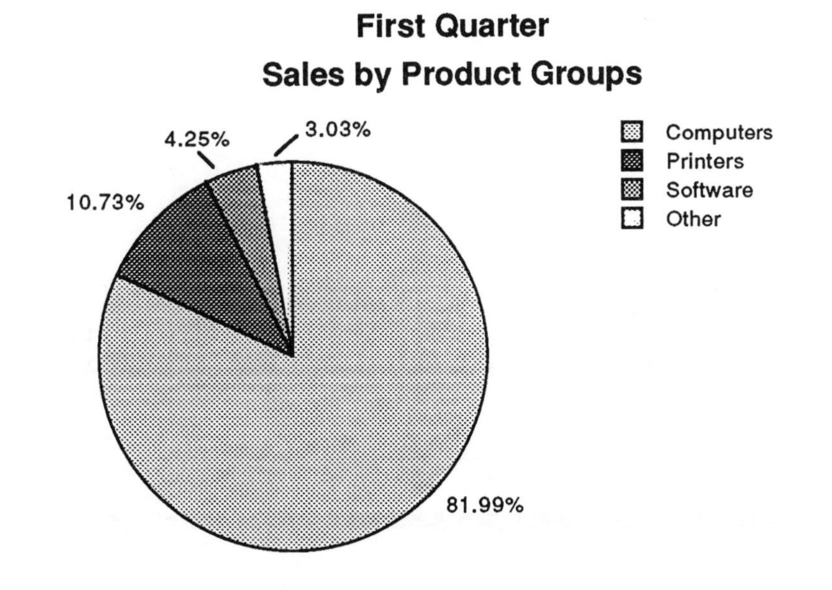

1. Which two product groups had nearly equal sales? ____________

2. Computer sales accounted for what percent of the total sales? ____________

3. About how many times greater were computer sales than printer sales? (Give the answer as a whole number.) ____________

4. Which product group accounted for the greater sales amount: software or other products? ____________

5. What product group is represented by the section of the circle shaded the darkest? ____________

Answers begin on page 257.

Exercise 4

Count On Us Computers does business with City Office Supply. The City Office Supply bar graph shown below shows sales in three cities during each month of the past year. The line graph shows total sales for all three cities combined.Study the graph and answer the questions below.

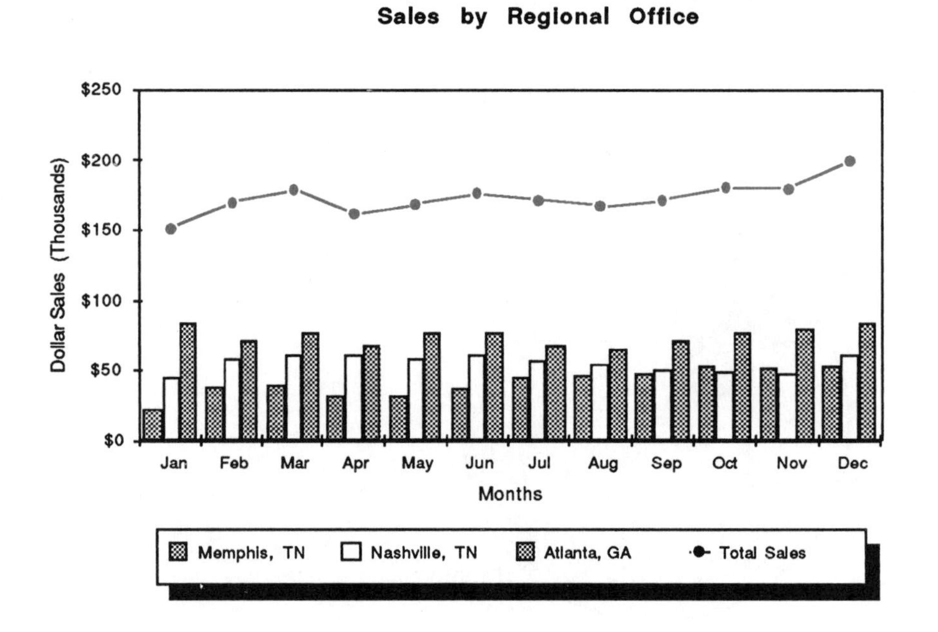

1. Based on the line graph, sales are showing __________
   a. a gradual decrease.
   b. a gradual increase.
   c. no change.
   d. a dramatic increase.

2. Which month had the greatest total sales? ______________________

3. Which month had the least total sales? ______________________

4. Which month had the smallest sales amount for Memphis? ______________________

5. Which of the three cities had the greatest sales amount for the year? ______________________

Check Yourself

Circle the letter of the correct response.

1. Don wanted to make a graph to show how the year's budget was divided. Which type of graph would be best?
   a. circle
   b. pictograph
   c. bar
   d. line

Use the following expense summary to answer question 2 below.

Count on Us Computers

Training Department

YEARLY EXPENSE SUMMARY

MONTH	PHONE	MEALS	TRAVEL	HOTEL	MISC.	MONTHLY TOTAL
JANUARY	206.80	447.90	402.20	723.30	46.70	
FEBRUARY	230.20	502.40	447.80	800.20	66.50	
MARCH	298.00	613.20	480.00	876.50	80.00	
1ST QUARTER TOTAL						

2. Rachel made a bar graph of the training department's expenses for March. Which of these statements would be true?

a. The bar for travel expenses would be the tallest.
b. The bar for miscellaneous expenses would be the shortest.
c. The bar for phone expenses would be about half the size of the bar for meal expenses.
d. The bar for hotel expenses would be half the size of the bar for travel expenses.

Use line graph below when answering questions 3 through 5.

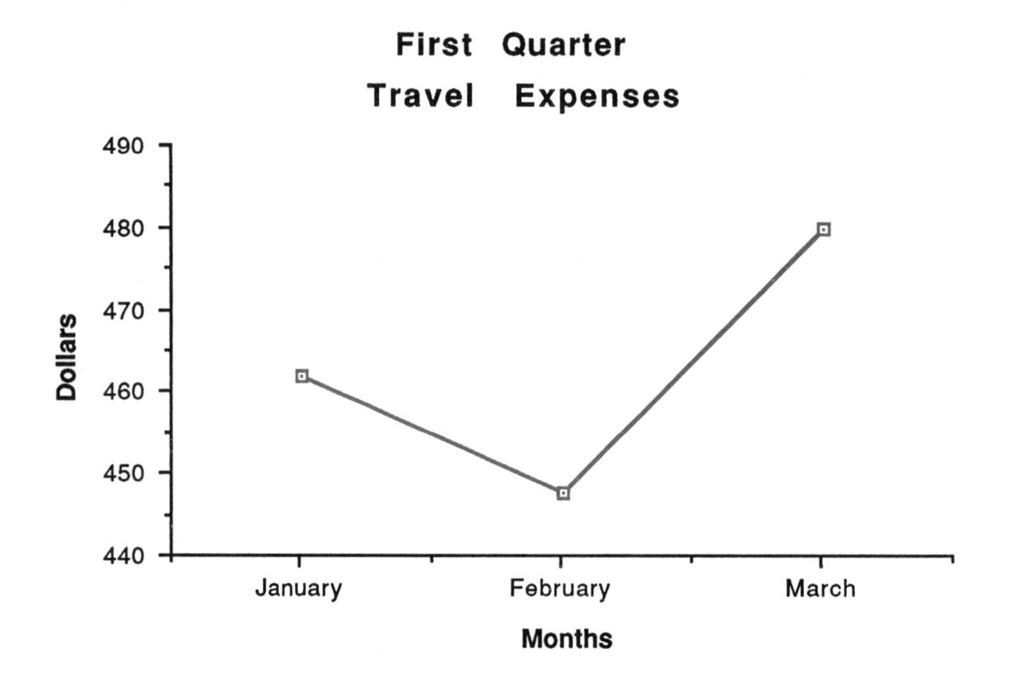

3. The vertical axis stands for _________

a. the name of the company.
b. the main idea of the graph.
c. the months of the first quarter.
d. dollar amounts.

4. The horizontal axis stands for _________

a. the name of the company.
b. the main idea of the graph.
c. the months of the first quarter.
d. dollar amounts.

5. Refer to the expense summary used in question 2 above. Is the line graph above a reasonable representation of those amounts?

a. Yes.
b. No, the point indicating travel expenses for January is too high.
c. No, the point indicating travel expenses for February is too low.
d. No, the point indicating travel expenses for March is too high.

Work Problem

A member of the training department created the graph below to show the training department's expenses for June. Verify that the graph is a reasonable representation of the expenses. Refer to the expense summary from question 2 above.

Is the graph reasonably correct? ____________________

If not, where is the mistake? ____________________

____

____

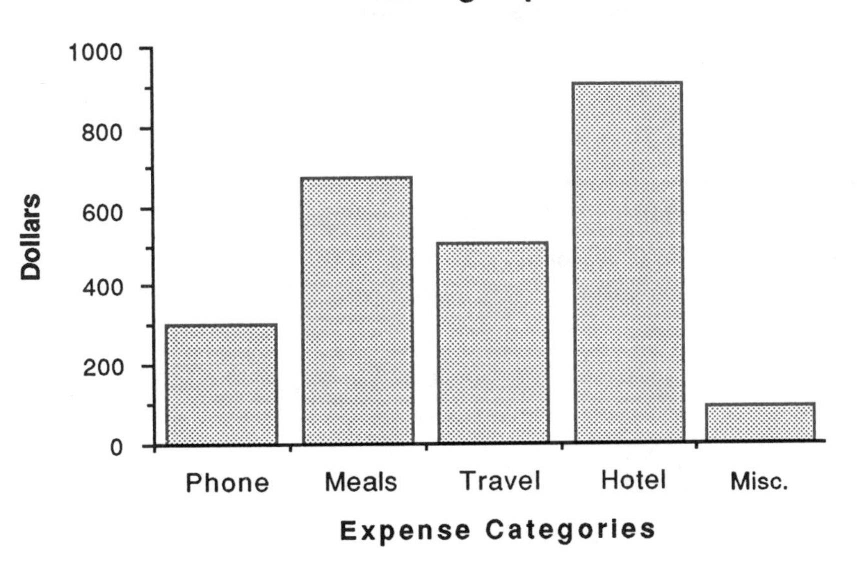

Answers to Problem-Solving Practice Questions

DEFINE the problem

- A verification of the expenses on the graph
- To verify the amounts

PLAN the solution

- The February expenses on the graph and the expense summary
- On the graph and the expense summary
- The February expenses on the graph and the expense summary
- No calculations will be necessary

SOLVE the problem

- The expenses shown on the graph
- No calculation is needed.
- No calculation is needed.
- Does not apply.
- The standards are the expenses for February: phone, $230.20; meals, $502.40; travel, $447.80; hotel, $800.20; miscellaneous, $66.50.
- The units are already the same.
- The hotel expenses were the greatest, so they should be the largest section of the circle. Meal and hotel expenses are nearly equal, with the section of the circle for meals being slightly larger. The section of the circle representing the phone expenses should be about half the size of the travel and hotel sections. The section for miscellaneous expenses should be the smallest. The circle graph looks correct.

CHECK the solution

- The defined purpose was accomplished. The graph was verified.
- The solution to the work problem is reasonable. The size of the sections of the pie graph are reasonable representations of the expenses.

Answers to Skills Practice Problems

Exercise 1

1. September
2. July
3. August, September and November
4. August
5. Months
6. Thousands of dollars
7. About $400,000

Exercise 2

1. April
2. July
3. July and August
4. May and June
5. There was a decrease in sales.
6. About $650,000

Exercise 3

1. Software and Other Products
2. 81.99%
3. 8 times
4. Software
5. Printers

LESSON 12

Verifying Data with a Table

Graphs are useful for showing relationships among amounts, but they are not a reliable source for exact amounts. When an exact amount is needed, refer to a table. Tables show many types of data, such as insurance rates, interest rates, or payment schedules. Payroll clerks like Curt Johnson depend on tax tables to verify the amount of federal income tax to deduct from employees' earnings.

Curt's solution to the following work problem is shown as an example.

Work Problem

Count On Us Computers keeps a record of each employee's earnings, deductions, and net pay. This record is called an *employee compensation record.* Before Curt added the deductions on Sharon's employee compensation record, he verified that the deduction for federal income tax was correct. He looked at the top of the record to determine Sharon's filing status and the number of withholding allowances. (The "M" after "filing status" stands for *married.* "S" stands for *single.*) Then, he referred to the tax tables on pages 259-260.

Employee Compensation Record

NAME: Sharon Goldman　　**RATE:** $8.15 per hour

SOCIAL SECURITY NUMBER: 395-74-6417　　**FILING STATUS:** M

WITHHOLDING ALLOWANCES: 1

	Hours		Earnings			Deductions				
Week	Reg	OT	Regular	Overtime	TOTAL	FICA	Fed Tax	State Tax	TOTAL	Net Pay
1/1	28	–	228.20	–	228.20	14.15	17.00	9.13		

SINGLE Persons—**WEEKLY** Payroll Period

And the wages are–		And the number of withholding allowances claimed is–										
At least	But less than	0	1	2	3	4	5	6	7	8	9	10
		The amount of income tax to be withheld shall be										
$0	$25	$0	$0	$0	$0	$0	$0	$0	$0	$0	$0	$0
25	30	1	0	0	0	0	0	0	0	0	0	0
30	35	1	0	0	0	0	0	0	0	0	0	0
35	40	2	0	0	0	0	0	0	0	0	0	0
40	45	3	0	0	0	0	0	0	0	0	0	0
45	50	4	0	0	0	0	0	0	0	0	0	0
50	55	4	0	0	0	0	0	0	0	0	0	0
55	60	5	0	0	0	0	0	0	0	0	0	0
60	65	6	0	0	0	0	0	0	0	0	0	0
65	70	7	0	0	0	0	0	0	0	0	0	0
70	75	7	1	0	0	0	0	0	0	0	0	0
75	80	8	2	0	0	0	0	0	0	0	0	0
80	85	9	3	0	0	0	0	0	0	0	0	0
85	90	10	3	0	0	0	0	0	0	0	0	0
90	95	10	4	0	0	0	0	0	0	0	0	0
95	100	11	5	0	0	0	0	0	0	0	0	0
100	105	12	6	0	0	0	0	0	0	0	0	0
105	110	13	6	0	0	0	0	0	0	0	0	0
110	115	13	7	1	0	0	0	0	0	0	0	0
115	120	14	8	2	0	0	0	0	0	0	0	0
120	125	15	9	2	0	0	0	0	0	0	0	0
125	130	16	9	3	0	0	0	0	0	0	0	0
130	135	16	10	4	0	0	0	0	0	0	0	0
135	140	17	11	5	0	0	0	0	0	0	0	0
140	145	18	12	5	0	0	0	0	0	0	0	0
145	150	19	12	6	0	0	0	0	0	0	0	0
150	155	19	13	7	1	0	0	0	0	0	0	0
155	160	20	14	8	1	0	0	0	0	0	0	0
160	165	21	15	8	2	0	0	0	0	0	0	0
165	170	22	15	9	3	0	0	0	0	0	0	0
170	175	22	16	10	4	0	0	0	0	0	0	0
175	180	23	17	11	4	0	0	0	0	0	0	0
180	185	24	18	11	5	0	0	0	0	0	0	0
185	190	25	18	12	6	0	0	0	0	0	0	0
190	195	25	19	13	7	0	0	0	0	0	0	0
195	200	26	20	14	7	1	0	0	0	0	0	0
200	210	27	21	15	9	2	0	0	0	0	0	0
210	220	29	22	16	10	4	0	0	0	0	0	0
220	230	30	24	18	12	5	0	0	0	0	0	0
230	240	32	25	19	13	7	1	0	0	0	0	0
240	250	33	27	21	15	8	2	0	0	0	0	0
250	260	35	28	22	16	10	4	0	0	0	0	0
260	270	36	30	24	18	11	5	0	0	0	0	0
270	280	38	31	25	19	13	7	0	0	0	0	0
280	290	39	33	27	21	14	8	2	0	0	0	0
290	300	41	34	28	22	16	10	3	0	0	0	0
300	310	42	36	30	24	17	11	5	0	0	0	0
310	320	44	37	31	25	19	13	6	0	0	0	0
320	330	45	39	33	27	20	14	8	2	0	0	0
330	340	47	40	34	28	22	16	9	3	0	0	0
340	350	48	42	36	30	23	17	11	5	0	0	0
350	360	50	43	37	31	25	19	12	6	0	0	0
360	370	51	45	39	33	26	20	14	8	2	0	0
370	380	53	46	40	34	28	22	15	9	3	0	0
380	390	54	48	42	36	29	23	17	11	5	0	0
390	400	56	49	43	37	31	25	18	12	6	0	0
400	410	57	51	45	39	32	26	20	14	8	1	0
410	420	59	52	46	40	34	28	21	15	9	3	0
420	430	61	54	48	42	35	29	23	17	11	4	0
430	440	64	55	49	43	37	31	24	18	12	6	0
440	450	67	57	51	45	38	32	26	20	14	7	1
450	460	70	58	52	46	40	34	27	21	15	9	3
460	470	73	61	54	48	41	35	29	23	17	10	4
470	480	75	64	55	49	43	37	30	24	18	12	6
480	490	78	67	57	51	44	38	32	26	20	13	7
490	500	81	69	58	52	46	40	33	27	21	15	9
500	510	84	72	61	54	47	41	35	29	23	16	10
510	520	87	75	63	55	49	43	36	30	24	18	12
520	530	89	78	66	57	50	44	38	32	26	19	13
530	540	92	81	69	58	52	46	39	33	27	21	15

MARRIED Persons—**WEEKLY** Payroll Period

And the wages are—		And the number of withholding allowances claimed is—										
At least	But less than	0	1	2	3	4	5	6	7	8	9	10
		The amount of income tax to be withheld shall be										
$0	$25	$0	$0	$0	$0	$0	$0	$0	$0	$0	$0	$0
25	75	1	0	0	0	0	0	0	0	0	0	0
75	80	1	0	0	0	0	0	0	0	0	0	0
80	85	2	0	0	0	0	0	0	0	0	0	0
85	90	3	0	0	0	0	0	0	0	0	0	0
90	95	4	0	0	0	0	0	0	0	0	0	0
95	100	4	0	0	0	0	0	0	0	0	0	0
100	105	5	0	0	0	0	0	0	0	0	0	0
105	110	6	0	0	0	0	0	0	0	0	0	0
110	115	7	0	0	0	0	0	0	0	0	0	0
115	120	7	1	0	0	0	0	0	0	0	0	0
120	125	8	2	0	0	0	0	0	0	0	0	0
125	130	9	3	0	0	0	0	0	0	0	0	0
130	135	10	3	0	0	0	0	0	0	0	0	0
135	140	10	4	0	0	0	0	0	0	0	0	0
140	145	11	5	0	0	0	0	0	0	0	0	0
145	150	12	6	0	0	0	0	0	0	0	0	0
150	155	13	6	0	0	0	0	0	0	0	0	0
155	160	13	7	1	0	0	0	0	0	0	0	0
160	165	14	8	2	0	0	0	0	0	0	0	0
165	170	15	9	2	0	0	0	0	0	0	0	0
170	175	16	9	3	0	0	0	0	0	0	0	0
175	180	16	10	4	0	0	0	0	0	0	0	0
180	185	17	11	5	0	0	0	0	0	0	0	0
185	190	18	12	5	0	0	0	0	0	0	0	0
190	195	19	12	6	0	0	0	0	0	0	0	0
195	200	19	13	7	1	0	0	0	0	0	0	0
200	210	21	14	8	2	0	0	0	0	0	0	0
210	220	22	16	10	3	0	0	0	0	0	0	0
220	230	24	17	11	5	0	0	0	0	0	0	0
230	240	25	19	13	6	0	0	0	0	0	0	0
240	250	27	20	14	8	2	0	0	0	0	0	0
250	260	28	22	16	9	3	0	0	0	0	0	0
260	270	30	23	17	11	5	0	0	0	0	0	0
270	280	31	25	19	12	6	0	0	0	0	0	0
280	290	33	26	20	14	8	2	0	0	0	0	0
290	300	34	28	22	15	9	3	0	0	0	0	0
300	310	36	29	23	17	11	5	0	0	0	0	0
310	320	37	31	25	18	12	6	0	0	0	0	0
320	330	39	32	26	20	14	8	1	0	0	0	0
330	340	40	34	28	21	15	9	3	0	0	0	0
340	350	42	35	29	23	17	11	4	0	0	0	0
350	360	43	37	31	24	18	12	6	0	0	0	0
360	370	45	38	32	26	20	14	7	1	0	0	0
370	380	46	40	34	27	21	15	9	3	0	0	0
380	390	48	41	35	29	23	17	10	4	0	0	0
390	400	49	43	37	30	24	18	12	6	0	0	0
400	410	51	44	38	32	26	20	13	7	1	0	0
410	420	52	46	40	33	27	21	15	9	2	0	0
420	430	54	47	41	35	29	23	16	10	4	0	0
430	440	55	49	43	36	30	24	18	12	5	0	0
440	450	57	50	44	38	32	26	19	13	7	1	0
450	460	58	52	46	39	33	27	21	15	8	2	0
460	470	60	53	47	41	35	29	22	16	10	4	0
470	480	61	55	49	42	36	30	24	18	11	5	0
480	490	63	56	50	44	38	32	25	19	13	7	0
490	500	64	58	52	45	39	33	27	21	14	8	2
500	510	66	59	53	47	41	35	28	22	16	10	3
510	520	67	61	55	48	42	36	30	24	17	11	5
520	530	69	62	56	50	44	38	31	25	19	13	6
530	540	70	64	58	51	45	39	33	27	20	14	8
540	550	72	65	59	53	47	41	34	28	22	16	9
550	560	73	67	61	54	48	42	36	30	23	17	11
560	570	75	68	62	56	50	44	37	31	25	19	12
570	580	76	70	64	57	51	45	39	33	26	20	14
580	590	78	71	65	59	53	47	40	34	28	22	15
590	600	79	73	67	60	54	48	42	36	29	23	17
600	610	81	74	68	62	56	50	43	37	31	25	18
610	620	82	76	70	63	57	51	45	39	32	26	20
620	630	84	77	71	65	59	53	46	40	34	28	21

Refer to the tax tables. Is Sharon's deduction for federal income tax correct?

Curt determined that the deduction was correct. Students who are unsure about using tables, might need more practice. The Skills Practice beginning on page 271 provides additional assistance and practice using tables.

Doing Math to Verify Data

Curt had defined his purpose as verifying Sharon's federal income tax deduction. He planned to compare the data on the employee compensation record to the standard on the tax table. He solved the problem, following the steps shown below.

1. **Select the relevant data.** *The federal tax amount listed on Sharon's employee compensation record: $17.00*

2. **Set up the calculation.** *No calculations are needed.*

3. **Do the calculation.** *No calculations are needed.*

4. **Check the accuracy of the answer.** *Does not apply.*

5. **Identify the standard.** *Curt looked on the tax table for married persons. Sharon earned $228.20, which falls between $220 and $230 on the table. Her withholding allowance, as shown on the employee compensation record, is 1. Curt located the intersection of the column for 1 withholding allowance and the row for wages between $220 and $230. He saw that $17.00 should be withheld.*

6. **Equate the units of measure.** *The units are already the same.*

7. Compare the amounts. *Since $17 equals $17, the amount on the employee compensation record was correct.*

Now an entire problem will be completed.

Work Problem

Curt is preparing the pay checks for the week of January 1. He must verify that the amount listed for Tim's federal income tax is correct.

Employee Compensation Record

NAME: Tim Holt

RATE: $8.15 per hour

SOCIAL SECURITY NUMBER: 123-45-6789

FILING STATUS: M

WITHHOLDING ALLOWANCES: 3

	Hours		Earnings			Deductions				
Week	Reg	OT	Regular	Overtime	TOTAL	FICA	Fed Tax	State Tax	TOTAL	Net Pay
1/1	32	–	260.80	–	260.80	16.17	11.00	10.43		

DEFINE the problem

- What is the expected outcome? *A verification of the amount deducted from Tim's earnings for federal income tax*
- What is the purpose? *To verify data*

PLAN the solution

- What data are needed? *The federal taxes written on the employee compensation record and the tax table*
- Where can the data be found? *On the employee compensation record and the tax table*
- What is already known? *Federal taxes from the employee compensation record and the tax table*
- Which operations should be used? *No calculation will be necessary*

SOLVE the problem

- Select the relevant data. *The federal tax amount listed on Tim's employee compensation record ($18.00)*
- Set up the calculation. *No calculation is needed*
- Do the calculation. *No calculation is needed*
- Check the accuracy of the answer. *Does not apply*

- Identify the standard.

Tim earned $260.80, which is between $260 and $270 on the tax table. His withholding allowance is 3. Curt located the intersection of the column for 3 withholding allowances and the row for wages between $260 and $270 on the tax table. He saw that $11.00 should be withheld.

- Equate the units of measure.

The units are already the same.

- Compare the amounts.

Since $11 differs from $18, there is a mistake on the employee compensation record.

CHECK the solution

Make sure the solution solves the work problem.

- Was the defined purpose accomplished?

Yes, the amount deducted from Tim's earnings for federal income tax was verified.

- Is the solution to the work problem reasonable?

Since $11 is not the same as $18, Curt looked on the table again to be certain that he used the correct table, the correct row and the correct column. Because Curt checked the table twice, the solution is reasonable

Note: When an amount disagrees with the standard, check the work to be sure. This ensures that the verification is reasonable.

Problem-Solving Practice

Use the DEFINE, PLAN, SOLVE, and CHECK steps to solve the next problem.

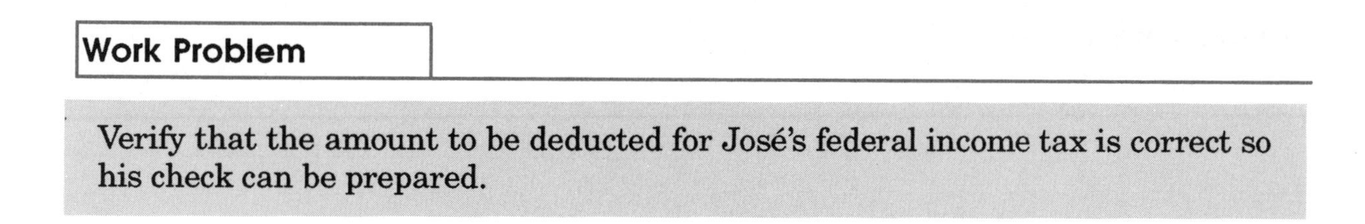

Work Problem

Verify that the amount to be deducted for José's federal income tax is correct so his check can be prepared.

Employee Compensation Record

NAME: José Montez

RATE: $493 per week

SOCIAL SECURITY NUMBER: 399-72-4684

FILING STATUS: M

WITHHOLDING ALLOWANCES: 4

	Hours		Earnings			Deductions				
Week	**Reg**	**OT**	**Regular**	**Overtime**	**TOTAL**	**FICA**	**Fed Tax**	**State Tax**	**TOTAL**	**Net Pay**
1/1	40	–	493.00	–	493.00	30.57	39.00	19.72		

DEFINE the problem

- What is the expected outcome? ______________________________

____

____

- What is the purpose? ______________________________

PLAN the solution

- What data are needed? ____________________
- Where can the data be found? ____________________
- What is already known? ____________________
- Which operations should be used? ____________________

SOLVE the problem

- Select the relevant data. ____________________
- Set up the calculation.
- Do the calculation.

- Check the accuracy of the answer.

- Identify the standard.

- Equate the units of measure.

- Compare the amounts.

CHECK the solution

Make sure the solution solves the work problem.

- Was the defined purpose accomplished?

- Is the solution reasonable?

Answers to the Problem-Solving Practice questions appear on page 277.

On Your Own

Solve the following work problems. Remember to DEFINE the problem, PLAN the solution, SOLVE the problem, and CHECK the solution.

Work Problem A

Rachel says, "The paychecks must be prepared. Verify that the amount to be deducted for Diane's federal income tax is correct. Make any necessary changes to the form, and initial it when it is right."

Employee Compensation Record

NAME: Diane Kruger

RATE: $8.67 per hour

SOCIAL SECURITY NUMBER: 425-09-1116

FILING STATUS: S

WITHHOLDING ALLOWANCES: 2

	Hours		Earnings			Deductions				
Week	Reg	OT	Regular	Overtime	TOTAL	FICA	Fed Tax	State Tax	TOTAL	Net Pay
1/1	40	–	346.80	–	346.80	21.50	36.00	13.87		

Work Problem B

Verify that the amount to be deducted for Rachel's federal income tax is correct so her check can be prepared. Make any necessary changes on the form.

Employee Compensation Record

NAME: Rachel Peters

RATE: $524 per week

SOCIAL SECURITY NUMBER: 687-96-9700

FILING STATUS: M

WITHHOLDING ALLOWANCES: 4

	Hours		Earnings			Deductions				
Week	**Reg**	**OT**	**Regular**	**Overtime**	**TOTAL**	**FICA**	**Fed Tax**	**State Tax**	**TOTAL**	**Net Pay**
1/1-	40	–	524.00	–	524.00	32.49	50.00	20.96		

Work Problem C

Verify that the amount to be deducted for Curt's federal income tax is correct so his total deductions can be calculated. Make any necessary changes on the form.

Employee Compensation Record

NAME: Curt Johnson

RATE: $8.52 per hour

SOCIAL SECURITY NUMBER: 112-55-9000

FILING STATUS: S

WITHHOLDING ALLOWANCES: 1

	Hours		Earnings			Deductions				
Week	**Reg**	**OT**	**Regular**	**Overtime**	**TOTAL**	**FICA**	**Fed Tax**	**State Tax**	**TOTAL**	**Net Pay**
1/1	20	–	170.40	–	170.40	10.56	22.00	6.81		

Work Problem D

Verify that the amount to be deducted for Maureen's federal income tax is correct so her check can be prepared. Make any necessary changes on the form.

Employee Compensation Record

NAME: Maureen Mahoney

RATE: $487 per week

SOCIAL SECURITY NUMBER: 318-64-3535

FILING STATUS: S

WITHHOLDING ALLOWANCES: 3

	Hours		Earnings			Deductions				Net
Week	Reg	OT	Regular	Overtime	TOTAL	FICA	Fed Tax	State Tax	TOTAL	Pay
1/1										

Work Problem E

Verify that the amount to be deducted for Mark's federal income tax is correct so his total deductions can be calculated. Make any necessary changes on the form.

Employee Compensation Record

NAME: Mark Schmidt

RATE: $8.49 per hour

SOCIAL SECURITY NUMBER: 498-11-9811

FILING STATUS: M

WITHHOLDING ALLOWANCES: 2

	Hours		Earnings			Deductions				Net
Week	Reg	OT	Regular	Overtime	TOTAL	FICA	Fed Tax	State Tax	TOTAL	Pay
1/1	40	–	338.60	–	338.60	21.06	34.00	13.58		

Skills Practice: Using Tables

A table lists data in rows and columns. *Rows* are horizontal and *columns* are vertical. The table below is used to determine taxes that property owners must pay. The table lists five properties together with their assessed values, tax rates, and tax amounts.

PROPERTY TAX COMPUTATIONS

Property Number	Assessed Value	Tax Rate	Tax Amount
1	$55,000	5.0%	$2,750
2	70,000	5.0%	3,500
3	60,000	4.5%	2,700
4	50,000	4.0%	2,000
5	75,000	5.5%	4,125

To identify the amounts, a table includes a main heading, column headings, and row headings. The main heading indicates the main content of the table. The column headings are the names above columns, such as "assessed value," "tax rate," and "tax amount." The row headings are listed in the left column. On this table, the row headings are the property numbers.

Use the row headings and column headings to locate information. For example, in order to determine the tax amount for property 3, look under the column labeled "tax amount" and across the row labeled "3." The row and column meet at $2,700. Therefore, the owner of property 3 must pay $2700.

Exercise 1

The table below shows payment schedules for a series of *mortgage loans*. Mortgage loans are usually taken to purchase a house, but may be taken to purchase a car, boat, or other item. Answer the questions which follow the table.

MORTGAGE PAYMENT SCHEDULE
30-Year Loans

Interest Rate:	9.00%	9.50%	10.00%	10.50%	11.00%	11.50%
Loan Amount	Monthly Payment	Monthly Payment	Monthly Payment	Monthly Payment	Monthly Payment	Monthly Payment
$1,000.00	$8.11	$8.47	$8.84	$9.21	$9.59	$9.96
$5,000.00	$40.56	$42.37	$44.20	$46.05	$47.93	$49.82
$10,000.00	$81.11	$84.73	$88.40	$92.11	$95.85	$99.64
$15,000.00	$121.67	$127.10	$132.60	$138.16	$143.78	$149.46
$20,000.00	$162.23	$169.47	$176.80	$184.21	$191.71	$199.27
$25,000.00	$202.78	$211.83	$221.00	$230.27	$239.63	$249.09
$30,000.00	$243.34	$254.20	$265.20	$276.32	$287.56	$298.91
$35,000.00	$283.90	$296.57	$309.40	$322.37	$335.49	$348.73
$50,000.00	$405.57	$423.67	$442.00	$460.54	$479.27	$498.18
$75,000.00	$608.35	$635.50	$663.00	$690.80	$718.90	$747.28
$80,000.00	$648.91	$677.87	$707.19	$736.86	$766.83	$797.09
$90,000.00	$730.02	$762.60	$795.59	$828.96	$862.68	$896.73
$100,000.00	$811.14	$847.34	$883.99	$921.07	$958.54	$996.37

1. What is the highest loan amount in the table? ________________

2. What is the lowest interest rate? ________________

3. What is the monthly payment for a loan of $75,000 at an interest rate of 10.5%? ________________

4. How much total payment will be required for a 30-year $80,000 loan at 10.5% interest? ________________

Answers begin on page 277.

Exercise 2

A minimum monthly payment schedule for typical credit card purchases is shown below. Review the table and answer the questions that follow it.

MINIMUM MONTHLY PAYMENT SCHEDULE

MINIMUM MONTHLY PAYMENT SCHEDULE. Your minimum monthly payment will change only at the time of a later advance.

BALANCE: (Immediately after latest advance)	REGULAR MINIMUM MONTHLY PAYMENT	BALANCE: (Immediately after latest advance)	REGULAR MINIMUM MONTHLY PAYMENT
$ .01 – 1,500.00	$ 50	$ 5,400.01 – 5,700.00	$ 140
1,500.01 – 1,800.00	55	5,700.01 – 6,000.00	145
1,800.01 – 2,000.00	60	6,000.01 – 6,300.00	150
2,000.01 – 2,150.00	65	6,300.01 – 6,500.00	155
2,150.01 – 2,300.00	70	6,500.01 – 6,700.00	160
2,300.01 – 2,450.00	75	6,700.01 – 6,900.00	165
2,450.01 – 2,600.00	80	6,900.01 – 7,200.00	170
2,600.01 – 2,750.00	85	7,200.01 – 7,400.00	175
2,750.01 – 2,900.00	90	7,400.01 – 7,700.00	180
2,900.01 – 3,300.00	95	7,700.01 – 8,000.00	185
3,300.01 – 3,600.00	100	8,000.01 – 8,300.00	190
3,600.01 – 3,900.00	105	8,300.01 – 8,600.00	195
3,900.01 – 4,200.00	110	8,600.01 – 8,900.00	200
4,200.01 – 4,440.00	115	8,900.01 – 9,100.00	205
4,440.01 – 4,600.00	120	9,100.01 – 9,400.00	210
4,600.01 – 4,800.00	125	9,400.01 – 9,800.00	215
4,800.01 – 5,100.00	130	9,800.01 – 10,200.00	220
5,100.01 – 5,400.00	135	10,200.01 – 11,000.00	225
		% OF BALANCE ROUNDED TO NEXT HIGHER $5.00.	
		$ 11,000.01 – and over	2.05%

1. If the monthly balance on the credit card is between $2,750.01 and $2,900.00, what is the minimum monthly payment that the credit card company will allow?

____

2. If the monthly balance on the credit card is between $4,200.01 and $4,440.00, what is the minimum monthly payment that the credit card company will allow?

____

3. If the monthly balance is $4,785.74, what is the minimum monthly payment that the credit card company will allow?

____

4. If the monthly balance is $1,285.78, what is the minimum monthly payment that the credit card company will allow?

____

5. If the monthly balance is $773.28, what is the minimum monthly payment that the credit card company will allow?

____

Answers begin on page 277.

Exercise 3

The Internal Revenue Service assesses an income tax based on each person's taxable income during the year. Review the tax table shown below and answer the questions that follow it.

IRS Tax Table—*Continued*

If line 37 (taxable income) is—		And you are—			
At least	But less than	Single	Married filing jointly *	Married filing separately	Head of a household
		Your tax is—			
23,000					
23,000	23,050	4,127	3,454	4,513	3,454
23,050	23,100	4,141	3,461	4,527	3,461
23,100	23,150	4,155	3,469	4,541	3,469
23,150	23,200	4,169	3,476	4,555	3,476
23,200	23,250	4,183	3,484	4,569	3,484
23,250	23,300	4,197	3,491	4,583	3,491
23,300	23,350	4,211	3,499	4,597	3,499
23,350	23,400	4,225	3,506	4,611	3,506
23,400	23,450	4,239	3,514	4,625	3,514
23,450	23,500	4,253	3,521	4,639	3,521
23,500	23,550	4,267	3,529	4,653	3,529
23,550	23,600	4,281	3,536	4,667	3,536
23,600	23,650	4,295	3,544	4,681	3,544
23,650	23,700	4,309	3,551	4,695	3,551
23,700	23,750	4,323	3,559	4,709	3,559
23,750	23,800	4,337	3,566	4,723	3,566
23,800	23,850	4,351	3,574	4,737	3,574
23,850	23,900	4,365	3,581	4,751	3,581
23,900	23,950	4,379	3,589	4,765	3,592
23,950	24,000	4,393	3,596	4,779	3,606

If line 37 (taxable income) is—		And you are—			
At least	But less than	Single	Married filing jointly *	Married filing separately	Head of a household
		Your tax is—			
26,000					
26,000	26,050	4,967	3,904	5,353	4,180
26,050	26,100	4,981	3,911	5,367	4,194
26,100	26,150	4,995	3,919	5,381	4,208
26,150	26,200	5,009	3,926	5,395	4,222
26,200	26,250	5,023	3,934	5,409	4,236
26,250	26,300	5,037	3,941	5,423	4,250
26,300	26,350	5,051	3,949	5,437	4,264
26,350	26,400	5,065	3,956	5,451	4,278
26,400	26,450	5,079	3,964	5,465	4,292
26,450	26,500	5,093	3,971	5,479	4,306
26,500	26,550	5,107	3,979	5,493	4,320
26,550	26,600	5,121	3,986	5,507	4,334
26,600	26,650	5,135	3,994	5,521	4,348
26,650	26,700	5,149	4,001	5,535	4,362
26,700	26,750	5,163	4,009	5,549	4,376
26,750	26,800	5,177	4,016	5,563	4,390
26,800	26,850	5,191	4,024	5,577	4,404
26,850	26,900	5,205	4,031	5,591	4,418
26,900	26,950	5,219	4,039	5,605	4,432
26,950	27,000	5,233	4,046	5,619	4,446

If line 37 (taxable income) is—		And you are—			
At least	But less than	Single	Married filing jointly *	Married filing separately	Head of a household
		Your tax is—			
29,000					
29,000	29,050	5,807	4,354	6,193	5,020
29,050	29,100	5,821	4,361	6,207	5,034
29,100	29,150	5,835	4,369	6,221	5,048
29,150	29,200	5,849	4,376	6,235	5,062
29,200	29,250	5,863	4,384	6,249	5,076
29,250	29,300	5,877	4,391	6,263	5,090
29,300	29,350	5,891	4,399	6,277	5,104
29,350	29,400	5,905	4,406	6,291	5,118
29,400	29,450	5,919	4,414	6,305	5,132
29,450	29,500	5,933	4,421	6,319	5,146
29,500	29,550	5,947	4,429	6,333	5,160
29,550	29,600	5,961	4,436	6,347	5,174
29,600	29,650	5,975	4,444	6,361	5,188
29,650	29,700	5,989	4,451	6,375	5,202
29,700	29,750	6,003	4,459	6,389	5,216
29,750	29,800	6,017	4,470	6,403	5,230
29,800	29,850	6,031	4,484	6,417	5,244
29,850	29,900	6,045	4,498	6,431	5,258
29,900	29,950	6,059	4,512	6,445	5,272
29,950	30,000	6,073	4,526	6,459	5,286

1. What is the tax due on taxable income between $26,500 and $26,550 for a single person?

____

2. What is the tax due on taxable income between $29,550 and $29,600 for a married person filing jointly?

____

3. What is the tax due on taxable income of $26,325 for a married person filing separately?

____

4. What is the tax due on taxable income of $29,485 for a head of a household?

____

5. What is the tax due on taxable income of $26,075 for a married person filing jointly?

____

Check Yourself

Refer to the tax tables on pages 259 and 260 to answer these questions. Circle the letter of the correct response.

1. To determine the tax for a single person who claims two withholding allowances and earns $304 a week, look in the column marked _______________________

   a. single.
   b. 1.
   c. 2.
   d. $304.

2. How much federal income tax would be withheld for a married person with a weekly income of $191?

   a. $19
   b. $12
   c. $6
   d. Not enough information is provided.

3. How much federal income tax would be withheld for a single person with an income between $250 and $260 who claims one deduction?
   a. $35
   b. $28
   c. $22
   d. Not enough information is provided.

4. If the number on the table disagrees with the data being verified, then ________
   a. check to be sure the correct table is being used.
   b. check to be sure the correct column is being used.
   c. check to be sure the correct row is being used.
   d. all of the above.

5. Sharon is married and claims one withholding allowance. She earned $203.75 last week. Her employee compensation record lists $14 as the amount to be deducted for federal income tax. Is that amount correct?
   a. Yes.
   b. No, it should be $13.
   c. No, it should be $16.
   d. No, it should be $21.

Work Problem

The end of the pay period is approaching, and Laurie's paycheck must be prepared. Verify that the amount to be deducted for federal income tax is correct. Make any necessary changes on the form, and initial the form when it is correct.

Employee Compensation Record

NAME: Laurie Klitzke **RATE:** $7.88 per hour

SOCIAL SECURITY NUMBER: 315-41-6912 **FILING STATUS:** M

WITHHOLDING ALLOWANCES: 0

	Hours		Earnings			Deductions				
Week	Reg	OT	Regular	Overtime	TOTAL	FICA	Fed Tax	State Tax	TOTAL	Net Pay
1/1	20	–	157.60	–	157.60	9.77	13.00	6.30		

Answers to Problem-Solving Practice Questions

DEFINE the problem

- A verification of the amount deducted from José's earnings for federal income tax
- To verify data

PLAN the solution

- The federal taxes listed on the employee compensation record and on the tax table
- On the employee compensation record and the tax table
- Federal taxes from the employee compensation record, total earnings, number of deductions, filing status
- No calculation will be necessary.

SOLVE the problem

- The federal tax amount listed on José's employee compensation record: $39.00
- No calculation is needed.
- No calculation is needed.
- Does not apply.
- José earned $493.00, which is between $490 and $500 on the tax table for married persons. His withholding allowance is four. Locate the intersection of the column for four withholding allowances and the row for wages between $490 and $500. The amount that should be withheld is $39.00.
- The units are already the same.
- The amounts are equal: $39 equals $39.

CHECK the solution

- The defined purpose was accomplished. The amount deducted from José's earnings for federal income tax was verified.
- The standard agrees with the amount being verified, so the solution is reasonable.

Answers to Skills Practice Problems

Exercise 1

1. $100,000 **2.** 9.00% **3.** $690.80 **4.** $26,526.96

Exercise 2

1. $90 **2.** $115 **3.** $125 **4.** $50 **5.** $ 50

Putting It All Together

Work Problem A

The expense summary of the training department for the months of January through July, 19—, is shown below. The training department was told to stay within a budget of $2300 during July. Verify that the department stayed within budget. Determine the difference between the actual expenses and the budgeted amount.

Did the training department stay within the budgeted amount? ____________

What was the difference? ____________

Count on Us Computers

Training Department

YEARLY EXPENSE SUMMARY

MONTH	PHONE	MEALS	TRAVEL	HOTEL	ENTERTAINMENT	MISC.	MONTHLY TOTAL
JANUARY	206.80	447.90	402.20	723.30	0	46.70	1826.90
FEBRUARY	230.20	502.40	447.80	800.20	0	66.50	2047.10
MARCH	298.00	613.20	480.00	876.50	27.00	80.00	2374.70
1ST QUARTER TOTAL	735.00	1563.15	1330.00	2400.00	27.00	193.20	6248.70
APRIL	212.20	495.90	414.00	719.00	0	50.10	1895.20
MAY	242.80	510.10	459.30	789.70	0	71.80	2073.70
JUNE	301.00	674.00	503.90	905.80	32.00	92.20	2508.90
2ND QUARTER TOTAL	760.00	1680.00	1377.20	2414.50	32.00	214.10	6477.80
JULY	282.30	561.90	452.00	736.20	30.00	75.70	

Work Problem B

One of the trainers made the graph below to show the department expenses for July. Use the figures in Work Problem A, and verify that the graph looks reasonably correct.

Is the graph reasonably correct? ______________________________

If not, where is the mistake? ______________________________

____

____

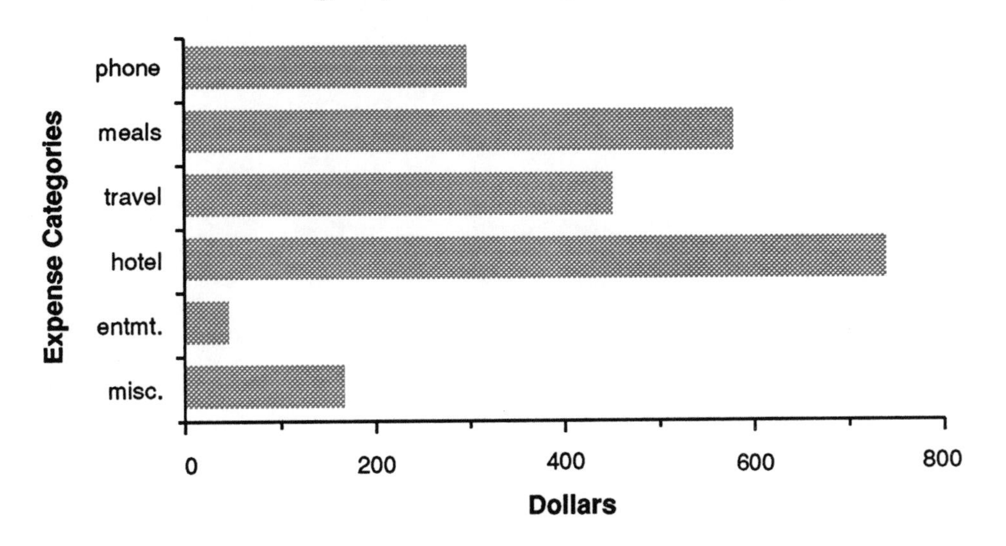

UNIT IV

ANALYZING AND INTERPRETING NUMBERS

Sometimes, office workers do more than find an amount or express a relationship; they *analyze* and *interpret* numbers. Analyzing and interpreting numbers requires thinking about numbers and how they compare, and then drawing a conclusion about them. For example, an employee who is asked to order enough computer paper for the next three months might want to look at previous months' paper use to see if the department is using more paper each month. He or she might look at the same months in the previous year to see if there is any unusual paper use in those months. Or the employee might determine an average month's paper use and decide to use that number to estimate how much paper to order. All of these are ways to analyze numbers. The process includes five steps:

1. Select the relevant data.
2. Set up the calculation.
3. Do the calculation.
4. Check the accuracy of the answer.
5. Draw the conclusion.

In this unit, students will analyze and interpret the numbers on sales summaries, supply summaries, inventories and resource tables. They practice sequencing numbers, finding averages, recognizing trends, and comparing percents.

Voyager Publishing

The setting for this unit is Voyager Publishing, a company which publishes travel books. The company is divided into four departments: sales and marketing, editor-ial, administration and accounting, and production. Stephanie Peña, Leonard Powell, Bonnie Pavek and Al Mancini work in the various departments.

LESSON 13

Analyzing and Interpreting Numbers by Sequencing Values

Sometimes office workers must do more than find an amount. They must *analyze* or *interpret* the amount to provide someone else with information. For example, a supervisor may be interested in knowing which product sold the best during a particular quarter of the year. In this situation, the office worker must rank the products according to how much of each was sold during the quarter.

Stephanie Peña's solution to the following work problem is shown as an example.

Work Problem

Voyager Publishing publishes several series of travel books, one series for each region of the world. A sales summary for the Asia series is shown below. Stephanie Peña is an office assistant at Voyager Publishing. Paul Yang, the president of the company, asked Stephanie, "Look at our sales of the Asia series for last quarter. Tell me which month had the highest unit sales."

This sales summary shows the number of *units* (books) that were sold. (Sales summaries sometimes show how many dollars were paid to the company for the units.)

VOYAGER PUBLISHING
SALES SUMMARY FOR THE ASIA SERIES—YEAR ONE

	BOOK TITLES				
	Inside India	Inside Japan	Inside China	Inside the Soviet Union	
MONTH	UNITS SOLD:				TOTALS
January	219	515	196	321	
February	283	640	272	489	
March	525	852	345	693	
1st Quarter					

First Stephanie added the monthly totals for each title. Add and write the answers on the sales summary above.

January ______________________

February ______________________

March ______________________

Stephanie arrived at these totals: January, 1251; February, 1684; and March, 2415. Then she compared the totals for each month and determined that March had the highest unit sales.

Students whose answers disagree with Stephanie's should refer to the Skills Practice section of Lesson 1 to review addition. The Skills Practice section on page 294 provides assistance and practice in comparing and interpreting results.

Doing Math to Analyze and Interpret Numbers

Paul Yang did not ask for the actual monthly totals. He wanted to know the month with the highest sales. Stephanie knew that she had to analyze and interpret the results of her calculations to give Paul the response he was expecting. Stephanie solved the problem by following the steps shown below.

Step			
1. Select the relevant data.	*The monthly totals for each book*		
2. Set up the calculations.	*January*	*February*	*March*
	219	283	525
	515	640	852
	196	272	345
	+ 321	+ 489	+ 693
3. Do the calculations.	1251	1684	2415
4. Check the accuracy of the answers.	321	489	693
	196	272	345
	515	640	852
	+ 219	+ 283	+ 525
	1251	1684	2415

5. Draw the conclusion.

The solution to the problem will not be the actual results of the calculations, but information based on the results. Stephanie compared the results: 1251, 1684, and 2415. The highest is 2415, which means that Voyager sold the most Asia series books in March.

Now Stephanie works through an entire problem, as shown below.

Work Problem

As president of the company, Paul likes to keep a watchful eye on sales totals. He said to Stephanie, "Find out which book in the Asia series sold the best during the first quarter."

DEFINE the problem

- What is the expected outcome? *A statement about which book sold best*
- What is the purpose? *To analyze and interpret numbers*

Note: Stephanie has been asked which book sold best. Words like best, highest, and lowest indicate that a *ranking* is necessary. Whenever data must be ranked, or put in numerical order, the purpose is to analyze and interpret numbers.

PLAN the solution

- What data are needed? *The number of books of each title sold during the quarter*
- Where can the data be found? *From calculations with the monthly totals on the sales summary*

- What is already known? *How many books of each title were sold each month*

- Which operation should be used? *Addition*

SOLVE the problem

- Select the relevant data. *Monthly totals for each book*

- Set up the calculations.

219	515	196	321
283	640	272	489
+ 525	+ 852	+ 345	+ 693

- Do the calculations.

1027	2007	813	1503

- Check the accuracy of the answers.

525	852	345	693
283	640	272	489
+ 219	+ 515	+ 196	+ 321
1027	2007	813	1503

- Draw the conclusion. *Inside Japan was the highest selling book.*

CHECK the solution

Make sure the solution solves the work problem.

- Was the defined purpose accomplished? *Yes, the results of the calculations were analyzed and interpreted. An answer to the president's question was found.*

- Is the solution to the work problem reasonable?

The sales totals for Inside Japan were consistently higher than those for the other books during each month of the first quarter. Therefore, it is reasonable that Inside Japan would have the highest total for the entire quarter.

Problem-Solving Practice

Use the DEFINE, PLAN, SOLVE, and CHECK steps to solve the following work problem.

Work Problem

Paul would like some information about sales in April, May, and June. What was the highest selling book during the second quarter? (Remember, the second quarter of the year is the second set of three months in the year—or April, May, and June.)

VOYAGER PUBLISHING
SALES SUMMARY FOR THE ASIA SERIES—YEAR ONE

	BOOK TITLES				
	Inside India	Inside Japan	Inside China	Inside the Soviet Union	
MONTH	**UNITS SOLD:**				**TOTALS**
January	219	515	196	321	1251
February	283	640	272	489	1684
March	525	852	345	693	2415
1st Quarter	1027	2007	813	1503	5350
April	519	962	447	924	
May	636	1153	598	1267	
June	844	1264	739	1425	
2nd Quarter					

DEFINE the problem

- What is the expected outcome? ______________________
- What is the purpose? ______________________

PLAN the solution

- What data are needed? ______________________
- Where can the data be found? ______________________
- What is already known? ______________________
- Which operation should be used? ______________________

SOLVE the problem

- Select the relevant data. ______________________
- Set up the calculations.
- Do the calculations.

- Check the accuracy of the answers.

- Draw the conclusion. ______________________________

CHECK the solution

Make sure the solution solves the work problem:

- Was the defined purpose accomplished? ______________________________

- Is the solution to the work problem reasonable? ______________________________

Answers to Problem-Solving Practice questions appear on page 301.

On Your Own

Solve the work problems given below. Remember to DEFINE the problem, PLAN the solution, SOLVE the problem, and CHECK the solution to make sure it solves the work problem.

Work Problem A

Paul would like more information about the second quarter. He asks, "In which month of the second quarter were sales the highest?" Use the information in the sales summary below to solve this problem.

Answer: ____________________

VOYAGER PUBLISHING SALES SUMMARY FOR THE ASIA SERIES—YEAR ONE					
	BOOK TITLES				
	Inside India	Inside Japan	Inside China	Inside the Soviet Union	
MONTH	**UNITS SOLD:**				**TOTALS**
January	219	515	196	321	1251
February	283	640	272	489	1684
March	525	852	345	693	2415
1st Quarter	1027	2007	813	1503	5350
April	519	962	447	924	
May	636	1153	598	1267	
June	844	1264	739	1425	
2nd Quarter					
July	708	1031	932	878	
August	512	846	1217	649	
September	311	629	675	518	
3rd Quarter					

Work Problem B

Gary Jacobs is the manager of the sales and marketing department. He must often prepare sales reports to present to Paul and the department managers. Gary asks, "Please analyze the unit sales results of each book for the third quarter. Then rank the books according to sales totals. List the book with the highest unit sales first." Use the sales summary on page 290 to solve this problem.

Answer:

1. ____________________

2. ____________________

3. ____________________

4. ____________________

Refer to the summary on page 293 to answer Work Problems C, D, and E.

Work Problem C

Gary would also like to know how the sales figures compared for each month of the third quarter. Rank the third quarter months according to unit sales. The month with the highest unit sales should be listed first.

Answer:

1. ____________________

2. ____________________

3. ____________________

Work Problem D

At the end of the fourth quarter, Gary would like the quarter's sales analyzed and interpreted. Compare the total sales for each book during the fourth quarter and rank the books from highest to lowest.

Answer:

1. ________________

2. ________________

3. ________________

4. ________________

Work Problem E

Gary is preparing the annual report. He asks, "During which quarter were sales the highest?" Find that information.

Answer: __________________

VOYAGER PUBLISHING SALES SUMMARY FOR THE ASIA SERIES—YEAR ONE					
	BOOK TITLES				
	Inside India	Inside Japan	Inside China	Inside the Soviet Union	
MONTH	**UNITS SOLD:**				**TOTALS**
January	219	515	196	321	1251
February	283	640	272	489	1684
March	525	852	345	693	2415
1st Quarter	1027	2007	813	1503	5350
April	519	962	447	924	
May	636	1153	598	1267	
June	844	1264	739	1425	
2nd Quarter					
July	708	1031	932	878	
August	512	846	1217	649	
September	311	629	675	518	
3rd Quarter					
October	277	743	628	471	
November	162	626	530	390	
December	110	531	393	319	
4th Quarter					
YEAR TOTALS					

Skills Practice: Ranking Data

Verifying data requires a comparison of the data to a standard. Interpreting data often requires a comparison among the data. This sometimes involves ranking or finding the highest or the lowest amount.

When comparing amounts, remember to consider the decimal point. Also, if amounts being compared are not expressed in the same unit, then one of the amounts must be converted.

Exercise 1

A. Circle the greatest amount in each line.

1.	3	1.3	13	.30	3.1
2.	.56	5	6	.65	5.06
3.	52.8	5.208	52.08	8.52	80.5
4.	8.27	8.72	8.07	8.2	8.702
5.	18 in.	2 ft.	20 in.	1 ft., 10 in.	
6.	48 oz.	1 lb., 8 oz.	2 lb.	15 oz.	

Answers begin on page 302.

B. Circle the smallest amount in each line.

1.	9.2	29	2.09	.902	2.29
2.	.666	6.06	.06	.066	6
3.	.703	3.1	1.7	1.37	7.13
4.	24.87	42.8	4.20	2.874	7.8
5.	19 days	2 wks., 6 days		30 days	3 weeks
6.	100 min.	1 hr., 15 min.		2 hrs.	1 1/2 hrs.

Answers begin on page 302.

Exercise 2

Office workers sometimes need to arrange forms, such as invoices and purchase orders, in numerical order. The invoice numbers listed below are out of order. Number them from 1 to 10, beginning with the lowest invoice number.

1.
______ 64047
______ 69314
______ 68814
______ 64039
______ 68813
______ 64023
______ 64048
______ 65246
______ 64124
______ 64034

2.
______ 24944
______ 23963
______ 25232
______ 25188
______ 23964
______ 24775
______ 24938
______ 24776
______ 24939
______ 23940

3.
______ 7007153
______ 7006019
______ 7006014
______ 7006090
______ 7006065
______ 7006015
______ 7006078
______ 7007081
______ 7006058
______ 7006023

4.
______ 8117112
______ 8118176
______ 8117113
______ 8117172
______ 8117168
______ 8117228
______ 8117315
______ 8117314
______ 8118142
______ 8118205

Answers begin on page 302.

Exercise 3

When office workers rank data, the ranking often proceeds from highest to lowest.

1. The sales summary below shows how many books were sold each month. Rank the months in order according to the number of books sold. The month with the greatest number of sales should be listed first. Write your answers on the lines.

VOYAGER PUBLISHING MONTHLY SALES SUMMARY	
MONTH	**BOOKS SOLD:**
January	3682
February	3625
March	2807
April	3089
May	3142
June	4418
July	3304
August	3499
September	2989
October	2506
November	4219
December	4488

1. December
2. ______________
3. ______________
4. ______________
5. ______________
6. ______________
7. ______________
8. ______________
9. ______________
10. ______________
11. ______________
12. ______________

Answers begin on page 302.

2. Voyager Publishing has twelve sales representatives. The sales summary shown below lists how many dollars worth of books each representative sold during the week of September 10. Rank the sales representatives from highest to lowest according to their sales dollars. Write your answers on the lines.

VOYAGER PUBLISHING SALES SUMMARY FOR REPRESENTATIVES	
SALES REP	**SALES IN DOLLARS**
Anderson	$4138.00
Chang	$1942.00
Davis	$3802.00
Foster	$2876.00
Graske	$4064.00
Haben	$3825.00
Judd	$2869.00
Knutson	$2996.00
Madden	$2777.00
Ness	$3577.00
Perez	$3810.00
Spangler	$4730.00

1. Spangler
2. ____________
3. ____________
4. ____________
5. ____________
6. ____________
7. ____________
8. ____________
9. ____________
10. ____________
11. ____________
12. ____________

Answers begin on page 302.

Exercise 4

A list of check numbers is given in sections A and B below. Determine which check numbers are missing. Write the missing numbers in order from lowest to highest on the lines at the right.

A. Check Numbers

Check Numbers	Missing Checks
12438	**1.** 12424
12432	**2.** ________
12433	**3.** ________
12422	**4.** ________
12442	**5.** ________
12428	**6.** ________
12437	**7.** ________
12423	**8.** ________
12441	**9.** ________
12434	**10.** ________
	11. ________
	12. ________

B. Check Numbers

Check Numbers	Missing Checks
21374	**1.** ________
21367	**2.** ________
21366	**3.** ________
21375	**4.** ________
21384	**5.** ________
21365	**6.** ________
21370	**7.** ________
21379	**8.** ________
21376	**9.** ________
21381	**10.** ________
	11. ________
	12. ________

Check Yourself

1. Which of these tasks would require an office worker to interpret results?
   a. Determining how many books are in inventory
   b. Determining how many books have been sold
   c. Determining the unit price of a book
   d. Determining which book sold the best during the year

2. When office workers interpret results, they ____________________
   a. determine a standard for the results.
   b. use the results to provide information.
   c. compare the results to a standard.
   d. equate the units of measure in the results.

3. Which of these steps is part of the strategy for interpreting results but not part of the strategy for finding amounts?
   a. Drawing the conclusion
   b. Checking the calculation
   c. Determining if the solution is reasonable
   d. Deciding which operation to use

Refer to the sales summary on the next page when answering questions 4 and 5 and completing the work problem.

VOYAGER PUBLISHING
SALES SUMMARY FOR THE ASIA SERIES—YEAR ONE

	BOOK TITLES				
	Inside India	Inside Japan	Inside China	Inside the Soviet Union	
MONTH	**UNITS SOLD:**				**TOTALS**
January	219	515	196	321	1251
February	283	640	272	489	1684
March	525	852	345	693	2415
1st Quarter	1027	2007	813	1503	5350
April	519	962	447	924	2852
May	636	1153	598	1267	3654
June	844	1264	739	1425	4272
2nd Quarter	1999	3379	1784	3616	10,778
July	708	1031	932	878	3549
August	512	846	1217	649	3224
September	311	629	675	518	2133
3rd Quarter	1531	2506	2824	2045	8906
October	277	743	628	471	2119
November	162	626	530	390	1708
December	110	531	393	319	1353
4th Quarter	549	1900	1551	1180	5180
YEAR TOTALS					

4. What was the best selling book in September?
   a. *Inside India*
   b. *Inside Japan*
   c. *Inside China*
   d. *Inside the Soviet Union*

5. In what month did *Inside India* sell the best?
   a. April
   b. May
   c. June
   d. July

Work Problem

Gary is working on his annual report. He asks, "What was the best selling book in the Asia series during the year?" Refer to the sales summary on page 300.

Answer: ____________________

Answers to Problem-Solving Practice Questions

DEFINE the problem

- A statement about which book sold the best
- To analyze and interpret numbers

PLAN the solution

- How many books of each title were sold during the quarter
- From calculations with the monthly totals on the sales summary
- How many books of each title were sold each month
- Addition

SOLVE the problem

- Monthly totals for each book

-

519	962	447	924
636	1153	598	1267
+ 844	+ 1264	+ 739	+ 1425

-

1999	3379	1784	3616

-

844	1264	739	1425
636	1153	598	1267
+ 519	+ 962	+ 447	+ 924
1999	3379	1784	3616

- *Inside the Soviet Union* was the best selling book.

CHECK the solution

- The purpose was accomplished. The results of the calculations were analyzed and interpreted.
- The solution is reasonable. The sales totals for *Inside the Soviet Union* were consistently higher than those for the other books during each month of the second quarter. Therefore, it is reasonable that *Inside the Soviet Union* would have the highest total for the entire quarter.

Answers to Skills Practice Problems

Exercise 1

A. **1.** 13 **2.** 6 **3.** 80.5 **4.** 8.72 **5.** 2 feet **6.** 48 ounces

B. **1.** .902 **2.** .06 **3.** .703 **4.** 2.874 **5.** 19 days **6.** 1 hr., 15 min.

Exercise 2

1. 4, 10, 9, 3, 8, 1, 5, 7, 6, 2
2. 8, 2, 10, 9, 3, 4, 6, 5, 7, 1
3. 10, 3, 1, 8, 6, 2, 7, 9, 5, 4
4. 1, 9, 2, 4, 3, 5, 7, 6, 8, 10

Exercise 3

1. **1.** December; **2.** June; **3.** November; **4.** January; **5.** February; **6.** August; **7.** July; **8.** May; **9.** April; **10.** September; **11.** March; **12.** October

2. **1.** Spangler; **2.** Anderson; **3.** Graske; **4.** Haben; **5.** Perez; **6.** Davis; **7.** Ness; **8.** Knutson; **9.** Foster; **10.** Judd; **11.** Madden; **12.** Chang

Office workers analyze and interpret averages whenever they compare averages to make a decision. When employees estimate how many supplies to order or how long a task will take to complete, they are often basing their estimate on averages. At Voyager Publishing, Leonard Powell is asked to analyze and interpret averages so that his supervisor can determine budget amounts.

Leonard's solution to the work problem is shown below.

Work Problem

Nancy Watson is the *controller* at Voyager Publishing. A controller is the person responsible for the company's finances. Nancy is Leonard's supervisor. At the end of the first quarter, Nancy asked Leonard, "Which department's average paper use for the quarter was the highest?"

VOYAGER PUBLISHING SUPPLY USAGE SUMMARY—YEAR ONE COPIER PAPER (reams)				
	Departments			
	Sales & Marketing	Editorial	Administration & Accounting	Company Totals
January	44	39	12	95
February	40	40	12	92
March	39	41	12	92
1st Qtr. Total				
1st Qtr. Avg.				

Leonard is responsible for the office supplies at Voyager Publishing. He keeps records of the amounts of supplies that are used. He uses these records to estimate how much to order for the future. (He also keeps the records so departments can be held accountable for their expenses.)

Leonard uses the supply usage summary shown above to keep an accurate record of the number of reams of copier paper each department uses each month. Each of the middle three columns of the summary represents a department in the company: sales and marketing, editorial, and administration and accounting.

Leonard totaled the monthly amounts used and determined the average, which he wrote on the row labeled "1st Quarter Avg." Then he compared the averages to find out which department's average was the highest. Use the supply summary above to calculate each department's average. Which department's monthly average was the highest?

Answer: ____________________

Leonard determined that sales and marketing used 123 reams of paper, for a monthly average of 41. The editorial department used 120 reams of paper, for an average of 40. The administration and accounting department used 36 reams of paper, for an average of 12. Sales and marketing had the highest monthly average. Students whose answers are different from Leonard's, or who need more practice determining averages, should turn to the Skills Practice section on page 316.

Doing Math to Analyze and Interpret Averages

Leonard planned to get the data he needed from the supply usage summary. Then he planned to add and divide to find the averages. He solved the problem following the steps shown below.

1. Select the relevant data. *The monthly totals for each department*

2. Set up the calculations.

44	39	12
40	40	12
+ 39	+ 41	+ 12

3. Do the calculations.

123   120   36

4. Check the accuracy of the answers.

39	41	12
40	40	12
+ 44	+ 39	+ 12
123	120	36

Leonard computed the totals. Then he needed to determine the averages. He followed the first four SOLVE steps again.

1. Select the relevant data.

The quarterly totals for each department and the number of monthly totals (3)

2. Set up the calculations.

$3\overline{)123}$ $3\overline{)120}$ $3\overline{)36}$

3. Do the calculations.

$$\begin{array}{r} 41 \\ 3\overline{)123} \\ 12 \\ \hline 3 \\ 3 \\ \hline 0 \end{array} \qquad \begin{array}{r} 40 \\ 3\overline{)120} \\ 12 \\ \hline 0 \end{array} \qquad \begin{array}{r} 12 \\ 3\overline{)36} \\ 3 \\ \hline 6 \\ 6 \\ \hline 0 \end{array}$$

4. Check the accuracy of the answers.

41	40	12
× 3	× 3	× 3
123	120	36

5. Draw the conclusion.

The highest number is 41. Sales and marketing had the highest monthly average.

Now an entire problem will be completed.

Work Problem

Nancy noticed that the company's photocopying expenses increased during the second quarter. She asked Leonard to determine which department had the highest monthly average use of copier paper during the second quarter.

VOYAGER PUBLISHING SUPPLY USAGE SUMMARY—YEAR ONE COPIER PAPER (reams)				
	Departments			
	Sales & Marketing	Editorial	Administration & Accounting	Company Totals
January	44	39	12	95
February	40	40	12	92
March	39	41	12	92
1st Qtr. Total				
1st Qtr. Avg.				
April	33	43	12	88
May	37	44	13	94
June	41	48	14	103
2nd Qtr. Total				
2nd Qtr. Avg.				

DEFINE the problem

- What is the expected outcome? *A statement indicating which department had the highest monthly average during the second quarter*

- What is the purpose? *To analyze and interpret averages*

PLAN the solution

- What data are needed? *The averages of each department's paper use during the second quarter*
- Where can the data be found? *From calculations with the monthly totals on the supply summary*
- What is already known? *How many reams of paper each department used each month*
- Which operations should be used? *Addition and division*

Note: Leonard plans to use addition and division because they are the operations necessary for finding averages.

SOLVE the problem

Note: Leonard will work through the first four SOLVE steps twice, once to add and once to divide.

- Select the relevant data. *Monthly totals for each department*
- Set up the calculations.

33	43	12
37	44	13
+ 41	+ 48	+ 14

- Do the calculations. 111 135 39

- Check the accuracy of the answers.

$$\begin{array}{r} 41 \\ 37 \\ +\ 33 \\ \hline 111 \end{array} \qquad \begin{array}{r} 48 \\ 44 \\ +\ 43 \\ \hline 135 \end{array} \qquad \begin{array}{r} 14 \\ 13 \\ +\ 12 \\ \hline 39 \end{array}$$

- Select the relevant data.

  Quarterly totals for each department and the number of monthly totals (3)

- Set up the calculations.

$$3\overline{)111} \qquad 3\overline{)135} \qquad 3\overline{)39}$$

- Do the calculations.

$$\begin{array}{r} 37 \\ 3\overline{)111} \\ \underline{9} \\ 21 \\ \underline{21} \\ 0 \end{array} \qquad \begin{array}{r} 45 \\ 3\overline{)135} \\ \underline{12} \\ 15 \\ \underline{15} \\ 0 \end{array} \qquad \begin{array}{r} 13 \\ 3\overline{)39} \\ \underline{3} \\ 9 \\ \underline{9} \\ 0 \end{array}$$

- Check the accuracy of the answers.

$$\begin{array}{r} 37 \\ \times\ 3 \\ \hline 111 \end{array} \qquad \begin{array}{r} 45 \\ \times\ 3 \\ \hline 135 \end{array} \qquad \begin{array}{r} 13 \\ \times\ 3 \\ \hline 39 \end{array}$$

- Draw the conclusion.

  Leonard compared the three averages. The editorial department had the highest monthly average during the second quarter.

CHECK the solution

Make sure the solution solves the work problem.

- Was the defined purpose accomplished?

  Yes, the results of the calculations were analyzed and interpreted.

- Is the solution to the work problem reasonable?

Yes, 37 is a reasonable average because it is between 33 and 41. Also, 45 is a reasonable average because it is between 43 and 48. Finally, 13 is a reasonable average because it is between 12 and 14. Leonard's comparison of the three averages (37, 45, and 13) was simple and straightforward. He also knows from experience that the editorial department makes many photocopies of large manuscripts. His conclusion that the editorial department had the highest average monthly paper use is reasonable.

Note: An average is reasonable if it is between the lowest and the highest amounts being added. For example, in this problem the sales and marketing department uses an average of 37 reams of copier paper a month. This is reasonable because 37 is between 33 (the amount for April) and 41 (the amount for June).

Problem-Solving Practice

Use the DEFINE, PLAN, SOLVE, and CHECK steps to solve the following work problem.

Work Problem

Nancy needs a way to reduce the photocopying expenses. She would like to know which department had the highest monthly average paper use during the third quarter. Solve this problem for Nancy.

VOYAGER PUBLISHING
SUPPLY USAGE SUMMARY—YEAR ONE
COPIER PAPER (reams)

	Departments			
	Sales & Marketing	Editorial	Administration & Accounting	Company Totals
January	44	39	12	95
February	40	40	12	92
March	39	41	12	92
1st Qtr. Total				
1st Qtr. Avg.				
April	33	43	12	88
May	37	44	13	94
June	41	48	14	103
2nd Qtr. Total				
2nd Qtr. Avg.				
July	40	57	12	109
August	41	49	13	103
September	42	44	14	100
3rd Qtr. Total				
3rd Qtr. Avg.				

DEFINE the problem

- What is the expected outcome? ______________________

- What is the purpose? ______________________

PLAN the solution

- What data are needed? ______________________

- Where can the data be found? ____________________

- What is already known? ____________________

- Which operations should be used? ____________________

SOLVE the problem

- Select the relevant data. ____________________

- Set up the calculations.

- Do the calculations.

- Check the accuracy of the answers.

- Do the first four steps again if needed.

- Draw the conclusion. ____________________

CHECK the solution

Make sure the solution solves the work problem:

- Was the defined purpose accomplished? ____
  ____
  ____

- Is the solution to the work problem reasonable? ____
  ____
  ____

Answers to Problem-Solving Practice questions appear on page 326.

On Your Own

Solve the work problems below. Remember to DEFINE the problem, PLAN the solution, SOLVE the problem, and CHECK the solution to make sure it solves the work problem. Refer to the supply usage summary shown below for work problems A, B and C.

VOYAGER PUBLISHING
SUPPLY USAGE SUMMARY—YEAR TWO
COPIER PAPER (reams)

	Departments			
	Sales & Marketing	Editorial	Administration & Accounting	Company Totals
January	44	39	12	95
February	40	40	12	92
March	39	41	12	92
1st Qtr. Total				
1st Qtr. Avg.				
April	33	43	12	88
May	37	44	13	94
June	41	48	14	103
2nd Qtr. Total				
2nd Qtr. Avg.				
July	40	57	12	109
August	41	49	13	103
September	42	44	14	100
3rd Qtr. Total				
3rd Qtr. Avg.				
October	41	44	13	98
November	44	39	15	98
December	44	43	14	101
4th Qtr. Total				
4th Qtr. Avg.				

Work Problem A

Nancy would like to know if Voyager Publishing used more paper as the year progressed. Find out if the company had a higher monthly average during the first quarter or the second quarter.

Answer: ____________________

Work Problem B

Nancy would like to check on the editorial department to see if they have reduced their use of copier paper. Was their average monthly use of paper lower in the first quarter or the third quarter?

Answer: ____________________

Work Problem C

Paul would like a comparison of the third and fourth quarters. During which of the two quarters did the company, as a whole, average the higher monthly use of paper?

Answer: ____________________

Use the following supply usage summary for Work Problems D and E.

Work Problem D

At the end of the following year, Gary, the manager of sales and marketing, reviewed paper usage during the year. He asked, "During which quarter was the paper use of the sales and marketing department at its lowest according to average monthly use?"

Answer: ____________________

VOYAGER PUBLISHING SUPPLY USAGE SUMMARY—YEAR TWO COPIER PAPER (reams)				
	Departments			
	Sales & Marketing	Editorial	Administration & Accounting	Company Totals
January	42	41	14	97
February	40	43	13	96
March	41	42	15	98
1st Qtr. Total				
1st Qtr. Avg.				
April	47	45	15	107
May	38	47	14	99
June	34	49	16	99
2nd Qtr. Total				
2nd Qtr. Avg.				
July	37	44	15	96
August	39	43	15	97
September	44	42	15	101
3rd Qtr. Total				
3rd Qtr. Avg.				
October	45	45	16	106
November	43	43	17	103
December	41	44	15	100
4th Qtr. Total				
4th Qtr. Avg.				

Work Problem E

Vanessa Harris, the manager of the editorial department, is preparing the budget for next year (Year Three). She needs to know the quarter during which the monthly average use of paper was highest for her department in Year Two. Find that information.

Answer: ____________________

Skills Practice: Comparing Averages

DETERMINING AN AVERAGE

When office workers refer to *averages*, they refer to a measure of a middle amount between the lowest and the highest possible amounts for a given situation. The average is calculated by adding all the amounts and then dividing the total by the number of amounts. Here is an example.

Part of Sandra's job is to verify orders from customers. Here is a list of days and the time she spent verifying orders each day.

DAY	TIME
Monday	50 minutes
Tuesday	40 minutes
Wednesday	55 minutes
Thursday	40 minutes
Friday	40 minutes

Sandra would like to know how much time she spends verifying orders on an average day.

Step 1. Add the amounts to determine a total. (In this case, add the number of minutes.)

$$\begin{array}{r} 50 \\ 40 \\ 55 \\ 40 \\ +\ 40 \\ \hline 225 \end{array}$$

Step 2. Count the amounts.

Five amounts of time are being added, one for each day.

Step 3. Divide the total by the count. (In this case, the total, 225, is divided by 5.)

$$\begin{array}{r} 45 \\ 5\overline{)225} \\ \underline{20} \\ 25 \\ \underline{25} \\ 0 \end{array}$$

Sandra verified orders for an average of 45 minutes each day.

Exercise 1

1. Refer to the partial expense summary for the editorial department, shown below, to answer the questions that follow.

VOYAGER PUBLISHING
EXPENSE SUMMARY

EDITORIAL DEPARTMENT

	Telephone	Postage	Office Supplies	Equipment Maintenance
January	$720	$430	$920	$260
February	830	510	1310	170
March	880	380	1430	320
April	770	480	1140	250
Totals				

a. What were the editorial department's average monthly postage expenses from January through April?

Answer: ______________

b. What were the editorial department's average monthly expenses for office supplies from January through April?

Answer: ____________________

c. What were the editorial department's average telephone expenses from January through April?

Answer: ____________________

d. What were the editorial department's average expenses for equipment maintenance from January through April?

Answer: ____________________

Answers begin on page 327.

2. Refer to the partial expense summary for the sales and marketing department, shown below, to answer the questions that follow.

VOYAGER PUBLISHING EXPENSE SUMMARY SALES AND MARKETING DEPARTMENT				
	Telephone	Postage	Office Supplies	Equipment Maintenance
January	$3100	$1000	$1100	$130
February	2500	1200	1000	140
March	3000	1600	800	150
April	2400	1800	900	150
May	3400	1300	1000	140
June	3600	1500	1200	190
Totals				

a. What were the sales and marketing department's average monthly postage expenses from January through June?

Answer: ____________________

b. What were the sales and marketing department's average monthly expenses for office supplies from January through June?

Answer: ____________________

c. What were the sales and marketing department's average monthly telephone expenses from January through June?

Answer: ____________________

d. What were the sales and marketing department's average monthly expenses for equipment maintenance from January through June?

Answer: ____________________

Answers begin on page 327.

DETERMINING MODE

An average is one kind of middle amount. Another kind of amount is the *mode*, or the amount which occurs most often. Sandra would like to know the time spent verifying orders last week that occurred most frequently—the mode.

DAY	TIME
Monday	50 minutes
Tuesday	40 minutes
Wednesday	55 minutes
Thursday	40 minutes
Friday	40 minutes

Compare the number of minutes for Monday through Sunday. Notice that 40 occurs 3 times, which is more often than either of the other amounts. So Sandra spent 40 minutes per day verifying orders—more than any other time period. In this case, the mode is 40 minutes.

Notice that this amount and the average are not the same. In this problem, the mode amount was 40 minutes, but the average was 45 minutes.

Exercise 2

1. Sandra recorded how much time it took her to type each of seven letters.

LETTER	TIME
Letter 1	20 minutes
Letter 2	15 minutes
Letter 3	20 minutes
Letter 4	15 minutes
Letter 5	20 minutes
Letter 6	25 minutes
Letter 7	25 minutes

What is the modal amount of time that it took Sandra to type a letter? ____________

2. Pao recorded how much time he spent filing invoices during each of eight days .

DATE	TIME
March 3	45 minutes
March 4	30 minutes
March 5	20 minutes
March 6	30 minutes
March 7	35 minutes
March 10	25 minutes
March 11	30 minutes
March 12	40 minutes

What is the mode time for filing invoices? ____________

3. Anita recorded the number of letters she typed per day during a two-week period.

DATE	LETTERS
March 17	4 letters
March 18	3 letters
March 19	5 letters
March 20	2 letters
March 21	3 letters
March 24	1 letters
March 25	3 letters
March 26	4 letters
March 27	3 letters
March 28	2 letters

What is the modal number of letters that Anita typed each day? ____________

4. Andrew is a switch board operator. He recorded the number of incoming calls during an eight-hour period.

PERIOD	CALLS
8:00 – 9:00	24 calls
9:00 – 10:00	12 calls
10:00 – 11:00	19 calls
11:00 – 12:00	14 calls
12:00 – 1:00	18 calls
1:00 – 2:00	19 calls
2:00 – 3:00	19 calls
3:00 – 4:00	21 calls

What is the modal number of calls received per hour? ____________

Answers begin on page 327.

Exercise 3

A partial expense summary for the administration and accounting department is shown below. Refer to it when answering the questions that follow.

VOYAGER PUBLISHING
EXPENSE SUMMARY

ADMINISTRATION AND ACCOUNTING DEPARTMENT

	Telephone	Postage	Office Supplies	Equipment Maintenance
January	429	124	275	35
February	521	202	275	80
March	521	124	311	35
April	521	141	275	40
May	479	124	322	35
June	535	185	366	35
Totals				

1. What were the department's average monthly postage expenses from January through April?

   Answer: ____________________

2. What is the mode for postage expenses each month?

   Answer: ____________________

3. What were the department's average monthly expenses for office supplies from January through April?

   Answer: ____________________

4. What is the mode for the amount spent on office supplies each month?

   Answer: ____________________

5. What were the department's average monthly telephone expenses from January through April?

   Answer: ____________________

6. What is the mode amount spent on telephone expenses each month?

   Answer: ____________________

7. What were the department's average monthly expenses for equipment maintenance from January through April?

   Answer: ____________________

8. What is the mode amount spent on equipment maintenance each month?

   Answer: ____________________

Check Yourself

Use the information below to answer questions 1 and 2.

Pao recorded how long it took him to drive to work each day during the week.

DAY	DRIVING TIME
Monday	32 minutes
Tuesday	18 minutes
Wednesday	18 minutes
Thursday	18 minutes
Friday	24 minutes

1. What was his average driving time?
- a. 18 minutes
- b. 22 minutes
- c. 24 minutes
- d. 32 minutes

2. What is the mode time required to get to work?
- a. 18 minutes
- b. 22 minutes
- c. 24 minutes
- d. 32 minutes

3. To determine an average, ______________________
- a. divide and then add.
- b. add and then divide.
- c. add and then multiply.
- d. divide and then subtract.

4. An average is reasonable when ______________________
- a. it is between the lowest and the highest amounts being added.
- b. it is the same as the number of amounts being added.
- c. it is smaller than the amounts being added.
- d. it is greater than the amounts being added.

5. When would an office worker most likely use an average?
- a. When determining how much money was spent on file folders
- b. When verifying the purchase order number for the file folders
- c. When determining how many file folders to order
- d. When determining the unit price for a file folder

Work Problem

Nancy Watson was preparing the budget for the coming year. She needs to know the Year Two quarter during which the average monthly use of paper was highest for the administration and accounting department. Find that information. Use the following supply usage summary.

Answer: ____________________

VOYAGER PUBLISHING
SUPPLY USAGE SUMMARY—YEAR TWO
COPIER PAPER (reams)

	Departments			
	Sales & Marketing	Editorial	Administration & Accounting	Company Totals
January	42	41	14	97
February	40	43	13	96
March	41	42	15	98
1st Qtr. Total				
1st Qtr. Avg.				
April	47	45	15	107
May	38	47	14	99
June	34	49	16	99
2nd Qtr. Total				
2nd Qtr. Avg.				
July	37	44	15	96
August	39	43	15	97
September	44	42	15	101
3rd Qtr. Total				
3rd Qtr. Avg.				
October	45	45	16	106
November	43	43	17	103
December	41	44	15	100
4th Qtr. Total				
4th Qtr. Avg.				

Answers to Problem-Solving Practice Questions

DEFINE the problem

- A statement indicating which department had the highest monthly average
- To analyze and interpret averages

PLAN the solution

- The month's averages of each department for the third quarter
- From calculations with the monthly totals on the supply summary
- How many reams of paper each department used each month during the third quarter
- Addition and division

SOLVE the problem

- Monthly totals for each department during the third quarter

- $\begin{array}{r} 40 \\ 41 \\ +\underline{42} \end{array}$ $\qquad$ $\begin{array}{r} 57 \\ 49 \\ +\underline{44} \end{array}$ $\qquad$ $\begin{array}{r} 12 \\ 13 \\ +\underline{14} \end{array}$

- 123 $\qquad$ 150 $\qquad$ 39

- $\begin{array}{r} 42 \\ 41 \\ +\underline{40} \\ 123 \end{array}$ $\qquad$ $\begin{array}{r} 44 \\ 49 \\ +\underline{57} \\ 150 \end{array}$ $\qquad$ $\begin{array}{r} 14 \\ 13 \\ +\underline{12} \\ 39 \end{array}$

- Quarterly total for each department and the number of monthly totals (3)

- $3\overline{)123}$ $\qquad$ $3\overline{)150}$ $\qquad$ $3\overline{)39}$

- $\begin{array}{r} 41 \\ 3\overline{)123} \\ \underline{12} \\ 3 \\ \underline{3} \\ 0 \end{array}$ $\qquad$ $\begin{array}{r} 50 \\ 3\overline{)150} \\ \underline{15} \\ 0 \end{array}$ $\qquad$ $\begin{array}{r} 13 \\ 3\overline{)39} \\ \underline{3} \\ 9 \\ \underline{9} \\ 0 \end{array}$

- $\begin{array}{r} 41 \\ \underline{\times 3} \\ 123 \end{array}$ $\qquad$ $\begin{array}{r} 50 \\ \underline{\times 3} \\ 150 \end{array}$ $\qquad$ $\begin{array}{r} 13 \\ \underline{\times 3} \\ 39 \end{array}$

- The editorial department had the highest monthly average during the third quarter.

CHECK the solution

- The purpose was accomplished. The results of the calculations were analyzed and interpreted. The highest average paper user was identified.
- The solution to the work problem is reasonable. Each of the averages is between the lowest and highest of the amounts added. Also, the conclusion is reasonable because Leonard knows that the editorial department makes photocopies of large book manuscripts.

Answers to Skills Practice Problems

Exercise 1

1.	**a.** $450	**b.** $1200	**c.** $800	**d.** $250
2.	**a.** 1400	**b.** $1000	**c.** $3000	**d.** $150

Exercise 2

1. 20 minutes **2.** 30 minutes **3.** 3 letters **4.** 19 phone calls

LESSON 15

Analyzing and Interpreting Trends

An office worker who is responsible for monitoring inventories or ordering supplies must be skilled at recognizing *trends*. A trend indicates the rate at which a series of amounts, such as monthly sales or supply use, is increasing or decreasing. A rapid change in a trend may signal a problem. For example, if the use of a product increases 10% each month for six months and then the use suddenly drops 50%, the situation might need investigation.

Bonnie Pavek solves the work problem as shown below.

Work Problem

The expense summary of the training department for the months of January through July is shown below. The training department was told to stay within a budget of $2300 during July. Verify that the department stayed within budget. Determine the difference between the actual expenses and the budgeted amount.

VOYAGER PUBLISHING
BOOK INVENTORY

Central and South America Series

TITLE	January 1	February 1	March 1	April 1
Inside Argentina	5000	4600	4100	3500
Inside Brazil	5000	4300	3700	3200
Inside Mexico	5000	4500	3800	2900

On the first day of each month, the number of books in stock is recorded on the inventory form shown above. To determine the sales for each month, Bonnie subtracts the end of the month figure from the beginning of the month figure.

On January 1, there were 5000 books on Argentina in inventory. At the end of the month, however, there were only 4600 books in inventory. Bonnie subtracted in order to determine that 400 books on Argentina were sold in January: 5000 − 4600 = 400. Bonnie subtracted each month from the preceding month to determine the sales figures.

January	5000	*February*	4600	*March*	4100
February	− 4600	*March*	− 4100	*April*	− 3500

Then she looked for a trend. Are sales of *Inside Argentina* increasing, decreasing or staying the same?

Answer: ____________________

If they are not staying the same, at what rate are they increasing or decreasing?

Answer: ____________________

Bonnie recognized a trend. She told Gary that the sales were increasing by 100 books each month. Students who did not come to the same conclusion as Bonnie should turn to the Skills Practice section on page 336 for more practice in recognizing trends.

Doing Math to Find, Analyze, and Interpret Trends

Bonnie planned to subtract the data on the inventory and then analyze and interpret the results in order to answer Gary's question.

1. **Select the relevant data.** *Number of books in stock each month*

2. **Set up the calculations.** *Bonnie compared each month's figures to the following month's figures in order to determine how the figures were changing. She wanted to know how the numbers were different, so she subtracted. (To find a difference, subtract.)*

5000	4600	4100
− 4600	− 4100	− 3500

3. Do the calculations.

400	500	600

4. Check the accuracy of the answers.

4600	4100	3500
+ 400	+ 500	+ 600
5000	4600	4100

5. Draw the conclusion.

Bonnie compared the sales totals to determine if they were increasing, decreasing, or staying the same. Because 600 is greater than 500 and 500 is greater than 400, she determined that they were increasing. Then she studied the totals more carefully to determine if there was a trend in the increase. She saw that the sales were increasing by 100 each month.

Now Bonnie works through an entire problem.

Work Problem

Vanessa edited *Inside Mexico*, so she is particularly interested in how well the book is selling. She called Bonnie and said, "Is there any trend that indicates how well *Inside Mexico* is selling? If so, what is the trend?"

VOYAGER PUBLISHING
BOOK INVENTORY

Central and South America Series

TITLE	January 1	February 1	March 1	April 1
Inside Argentina	5000	4600	4100	3500
Inside Brazil	5000	4300	3700	3200
Inside Mexico	5000	4500	3800	2900

DEFINE the problem

- What is the expected outcome? *A statement about the sales trend for Inside Mexico*

- What is the purpose? *To analyze and interpret a trend*

PLAN the solution

- What data are needed? *The changes in the inventory from month to month*

- Where can the data be found? *From calculations with the data on the inventory*

- What is already known? *How many books were in stock each month*

- Which operation should be used? *Subtraction*

SOLVE the problem

- Select the relevant data. *The number of books which were in stock each month*

- Set up the calculations.

$$\begin{array}{r} 5000 \\ -\,4500 \\ \hline \end{array} \qquad \begin{array}{r} 4500 \\ -\,3800 \\ \hline \end{array} \qquad \begin{array}{r} 3800 \\ -\,2900 \\ \hline \end{array}$$

- Do the calculations. 500 700 900

- Check the accuracy of the answers.

$$\begin{array}{r} 4500 \\ +\ 500 \\ \hline 5000 \end{array} \qquad \begin{array}{r} 3800 \\ +\ 700 \\ \hline 4500 \end{array} \qquad \begin{array}{r} 2900 \\ +\ 900 \\ \hline 3800 \end{array}$$

- Draw the conclusion.

  The sales figures for Inside Mexico are increasing by 200 each month.

CHECK the solution

Make sure the solution solves the work problem:

- Was the defined purpose accomplished?

  Yes, the results of the calculations were analyzed and interpreted. The trend was described.

- Is the solution to the work problem reasonable?

  The differences between the monthly totals on the inventory become larger. Therefore, it is reasonable to conclude that the sales figures for Inside Mexico are increasing.

Problem-Solving Practice

Use the DEFINE, PLAN, SOLVE, and CHECK steps to solve the following work problem.

Work Problem

Paul is studying the sales figures for various books. He would like to know if the sales figures for *Inside Brazil* follow a particular trend. Find that information.

VOYAGER PUBLISHING
BOOK INVENTORY

Central and South America Series

TITLE	January 1	February 1	March 1	April 1
Inside Argentina	5000	4600	4200	3800
Inside Brazil	5000	4300	3700	3200
Inside Mexico	5000	4500	3800	2900

DEFINE the problem

- What is the expected outcome? ______________________________
- What is the purpose? ______________________________

PLAN the solution

- What data are needed? ______________________________
- Where can the data be found? ______________________________
- What is already known? ______________________________
- Which operations should be used? ______________________________

SOLVE the problem

- Select the relevant data. ______________________________

- **Set up the calculations.**

- **Do the calculations.**

- **Check the accuracy of the answers.**

- **Draw the conclusion.** ______________________________

CHECK the solution

Make sure the solution solves the work problem:

- **Was the defined purpose accomplished?** ______________________________

____

____

- **Is the solution to the work problem reasonable?** ______________________________

____

____

Answers to Problem-Solving Practice questions appear on page 342.

On Your Own

Solve the following work problems. Refer to the book inventory form shown below to solve all five problems.

VOYAGER PUBLISHING BOOK INVENTORY Europe Series				
TITLE	**January 1**	**February 1**	**March 1**	**April 1**
Inside France	6900	6000	5200	4500
Inside Germany	7100	6300	5500	3000
Inside Italy	6500	5900	5100	4100
Inside Portugal	5400	5100	4800	4500
Inside Spain	5900	5250	4500	3650

Work Problem A

Gary is writing a sales report and would like to know if the sales figures for *Inside Italy* follow a particular trend. If so, he would like to know what the trend is. Determine that information.

Answer: ____________________

Work Problem B

Vanessa edited *Inside Portugal*, so she is particularly interested in how well the book is selling. What kind of trend do the sales of *Inside Portugal* show?

Answer: ____________________

Work Problem C

Vanessa would also like to know if the sales figures for *Inside Spain* are following the same trend as the sales figures for *Inside Portugal*. What kind of trend is indicated by the sales figures for *Inside Spain*?

Answer: ______________________________

Work Problem D

The author of *Inside France* receives royalties, which means she receives money for each book sold. The author would like to know what the sales trend is for *Inside France*. Determine that information.

Answer: ______________________________

Work Problem E

Due to political changes in Germany, Paul expects more people to travel to that country. Which month reflects a dramatic increase in the sales of *Inside Germany*?

Answer: ______________________________

Skills Practice: Trends

RECOGNIZING TRENDS

A *trend* is a pattern in the data which indicates what may happen to a company's future sales, income, and expenses. Employees of a company must be able to recognize trends in order to detect problems and make sound decisions about budgets and other business matters.

To identify a trend, identify the pattern in the data. Here is an example.

The data below shows how many rolls of stamps have been used in the accounting office at Voyager Publishing so far this year. What is the pattern in this data?

MONTH	ROLLS
January	9 rolls
February	10 rolls
March	11 rolls
April	12 rolls

Step 1. Set up the data.

Often it helps to set the data up horizontally, with space in the middle.

9 10 11 12

Step 2. Look at the first two numbers. Ask, "How are the numbers related?" Determine a rule that shows how the numbers relate. If the numbers are increasing, try a rule with addition or multiplication. If the numbers are decreasing, try a rule with subtraction or division.

In this example, 10 is 1 more than 9, so a good guess for the rule is "+ 1."

Data: 9 ↘ ↗ 10 11 12
Rule: +1

Step 3. Test the rule on the other numbers to be sure it is correct.

Data: 9 ↘ ↗ 10 ↘ ↗ 11 ↘ ↗ 12
Rule: +1 +1 +1

The rule describes the pattern. In this example, each succeeding number increases by one. Each month, one more roll of stamps is used.

Here is a more complicated example in which the rule itself has a pattern. Determine the rule for this series of numbers: 5, 9, 6, 10, 7

Step 1. Set up the data.

5 9 6 10 7

Step 2. Look at the first two numbers. Determine a rule that shows how the numbers relate.

In this example, 9 is 4 greater than 5, so a good guess for the rule is "+ 4."

Data: 5 9 6 10 7
Rule: ↘ +4 ↗

Step 3. Test the rule on the next two numbers. If it does not work, add to the rule to form a new two-part rule.

$9 + 4$ does not equal 6. However, $9 - 3 = 6$.

Data: 5 9 6 10 7
Rule: ↘ +4 ↗ ↘ −3 ↗

Step 4. Test the new two-part rule on the next numbers.

Data: 5 9 6 10 7
Rule: ↘ +4 ↗ ↘ −3 ↗ ↘ +4 ↗ ↘ −3 ↗

The rule is add 4 and subtract 3. The pattern is that each successive number increases by 4 and decreases by 3.

Exercise 1

Identify the rules in each of these number series.

1. 1 3 5 7 9

Rule: ______________________________

2. 30 27 24 21 18

Rule: ______________________________

3. 2 4 8 16 32

Rule: ______________________________

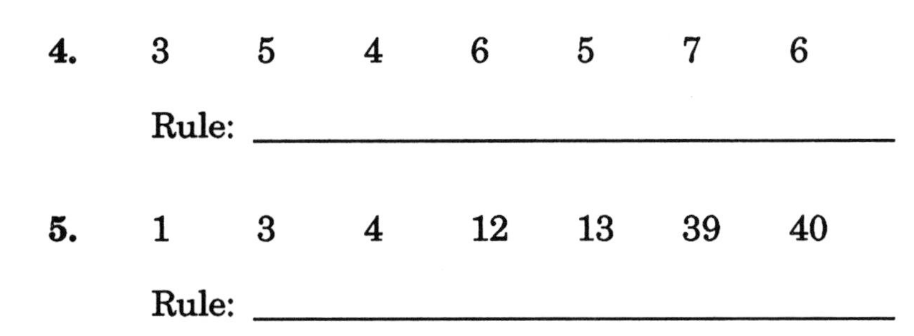

4. 3 5 4 6 5 7 6

Rule: ______________________________

5. 1 3 4 12 13 39 40

Rule: ______________________________

Answers begin on page 343.

Once the rule has been identified, it can be used to predict the next number in a series. As an example, find the next number in this series: 80, 40, 20, 10.

Step 1. Set up the data.

Data: 80 ↘ 40 ↘ 20 10
Rule: ÷ 2 ↗ ÷ 2 ↗

Step 2. Look at the first two numbers. Determine a rule which shows how the numbers relate. Sometimes more than one rule will work.

Data: 80 40 20 10
Rule: ÷2
Rule: −40

Step 3. Test both rules on the next number. Eliminate the one that does not work.

Data: 80 40 20 10
Rule: ÷2 ÷2
Rule: −40

Step 4. Test the rule on the next numbers.

Data: 80 40 20 10
Rule: ÷ 2 ÷ 2 ÷ 2

Step 5. Use the rule to predict the next number in the series.

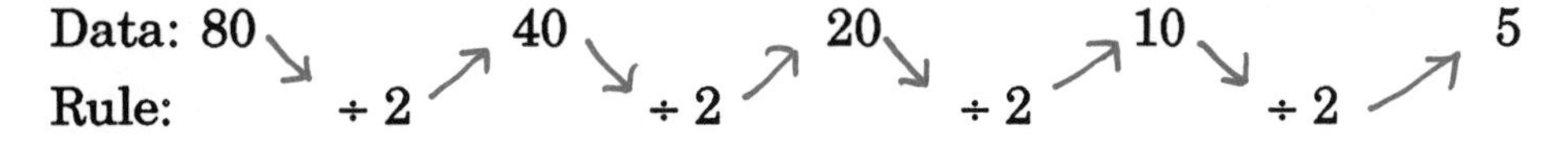

The next number in the series is 5.

Exercise 2

Find the next number in these series. Write the number on the line.

1. 5 10 15 20 ________
2. 59 53 47 41 ________
3. 21 28 35 42 ________
4. 3 9 27 81 ________
5. 90 70 80 60 ________

Answers begin on page 343.

Exercise 3

Use number patterns to determine when the quantity of each of the supplies listed on this inventory will reach zero. Put a zero (0) in the correct box.

OFFICE SUPPLY INVENTORY

ITEM	Jan 1	Feb 1	Mar 1	Apr 1	May 1	June 1	July 1
boxes of fax paper	60	48	36				
boxes of pens	90	75	60				
memo pads	150	125	100				
boxes of pencils	60	40	20				
computer disks	500	375	250				

Check Yourself

1. A trend _________
   a. is a pattern in the data.
   b. indicates what may happen to future sales, income, and expenses.
   c. is used to make business decisions.
   d. all of the above.

2. Who should be skilled at recognizing trends?
   a. Employees who monitor and order supplies
   b. Employees who monitor inventories
   c. Employees who plan budgets
   d. All of the above

3. Which number series shows a consistent pattern?
   a. 25, 50, 75
   b. 9, 20, 50
   c. 25, 9, 75
   d. None of the above.

Refer to the inventory shown below when answering questions 4 and 5 and the work problem.

VOYAGER PUBLISHING BOOK INVENTORY Mideast Series				
TITLE	**September 1**	**October 1**	**November 1**	**December 1**
Inside Egypt	3600	3200	2900	2700
Inside Israel	3400	3100	2900	2700
Inside Turkey	3900	3650	3450	3300

4. How many books about Turkey were sold in October?
   a. 200
   b. 250
   c. 3450
   d. 3650

5. In November, the sales of *Inside Turkey* ________________

a. increased.

b. decreased.

c. stayed the same.

Work Problem

Gary would like to know if political unrest in the Mideast is affecting the sales of *Inside Egypt*. What kind of trend do the sales figures indicate?

Answer: ____

Answers to Problem-Solving Practice Questions

DEFINE the problem

- A statement about the sales trend for *Inside Brazil*
- To analyze and interpret numbers

PLAN the solution

- The changes in the inventory from month to month
- From calculations with the data on the inventory
- How many books were in stock each month
- Subtraction

SOLVE the problem

- The number of books which were in stock each month
- $5000 - 4300$ $4300 - 3700$ $3700 - 3200$
- 700   600   500
- $4300 + 700 = 5000$ $3700 + 600 = 4300$ $3200 + 500 = 3700$
- The sales figures for *Inside Brazil* are decreasing by 100 each month.

CHECK the solution

- The defined purpose was accomplished. The results of the calculations were analyzed and interpreted. A statement was made about the trend.
- The solution to the work problem is reasonable. The differences between the months on the inventory are the same for each month. Therefore, it is reasonable to conclude that the sales figures for *Inside Brazil* are decreasing.

Answers to Skills Practice Problems

Exercise 1

1. add 2 **2.** subtract 3 **3.** multiply by 2 **4.** add 2 and subtract 1
5. multiply by 3 and add 1

Exercise 2

1. 25 **2.** 35 **3.** 49 **4.** 243 **5.** 70

Employees in marketing departments analyze and interpret data so they can learn more about potential customers and their spending habits. The data are sometimes presented as percentages, so office workers must be able to interpret percentages. Al Mancini is a marketing assistant in the sales and marketing department at Voyager Publishing. Part of his job is to help Gary, the manager of the department, analyze data about travelers.

Al's solution to the work problem is shown below.

Work Problem

The marketing department at Voyager Publishing uses resources such as the ones shown on the next page to learn more about where people are traveling overseas. This knowledge helps them focus their selling of books. Gary is analyzing the travel market for Asia. He asks Al to determine if more travelers went to Asia in Year 1 or Year 2.

U.S. Travelers to Foreign Countries
(excluding Canada and Mexico)
(in thousands)

	Year 1	Year 2	Year 3	Year 4	Year 5	Year 6
Total Overseas Travelers	12,290	11,560	11,810	12,310	13,310	12,520
Europe & Mediterranean	52%	46%	47%	45%	44%	45%
Caribbean & Cent. Amer.	28%	33%	31%	30%	31%	32%
South America	7%	6%	8%	8%	7%	9%
Asia	6%	8%	6%	9%	10%	8%
Africa	4%	3%	2%	3%	3%	3%
Australia	2%	2%	3%	2%	3%	2%
Other	1%	1%	1%	1%	1%	1%

Al needed to determine 6% of 12,290 and 8% of 11,560. Use the space below to calculate the answers.

Al determined that 6% of 12,290 is 737.4 and 8% of 11,560 is 924.8. More travelers went to Asia in Year 2 because 924.8 is larger than 737.4. Students whose answers disagree with Al's should turn to the Skills Practice section on page 354 for more practice with comparing percentages.

Doing Math to Analyze and Interpret Percentages

Al could not compare 6% to 8% because they represent parts of different totals. The 6% stands for 6% of 12,290 and the 8% stands for 8% of 11,560. Two percents cannot be directly compared unless they represent parts of the same number. Al planned to multiply to determine the amounts represented by each percent and then compare those two amounts to draw a conclusion.

He solved the problem by following the steps shown below.

1. **Select the relevant data.**

   The percent of travelers who went to Asia in Year 1 and in Year 2, and the total number of travelers for Year 1 and Year 2

2. **Set up the calculations.**

   $$\begin{array}{r} 12{,}290 \\ \times \quad .06 \\ \hline \end{array} \qquad \begin{array}{r} 11{,}560 \\ \times \quad .08 \\ \hline \end{array}$$

3. **Do the calculations.**

   737.4   924.8

4. **Check the accuracy of the answers.**

   $$\begin{array}{r} .06 \\ 12290\overline{)737.40} \\ 737.40 \\ \hline 0 \end{array} \qquad \begin{array}{r} .08 \\ 11560\overline{)924.80} \\ 925.80 \\ \hline 0 \end{array}$$

5. **Draw the conclusion.**

   Since 924.8 is greater than 737.4, more people traveled to Asia in Year 2 than in Year 1.

Note: Remember that the number of travelers on the table is listed in thousands, so 924.8 stands for 924,800 travelers and 737.4 stands for 737,400 travelers.

Now Al works through an entire problem.

Work Problem

Gary needs some other figures for Year 1 and Year 2. He asks Al, "Did more people travel to the Caribbean and Central America during Year 1 or during Year 2?"

DEFINE the problem

- What is the expected outcome? *A statement about which year more people traveled to the Caribbean and Central America*
- What is the purpose? *To analyze and interpret percentages*

PLAN the solution

- What data are needed? *How many travelers went to the Caribbean and Central America during Year 1 and Year 2*
- Where can the data be found? *From calculations with the percents on the table*
- What is already known? *The percent of travelers who went to the Caribbean and Central America during Year 1 and Year 2, and the total number of travelers for Year 1 and Year 2*
- Which operation should be used? *Multiplication*

SOLVE the problem

- Select the relevant data.

  The percent of travelers who went to the Caribbean and Central America during Year 1 and Year 2, and the total number of travelers for Year 1 and Year 2

- Set up the calculations.

$$\begin{array}{r} 12{,}290 \\ \times \quad .28 \\ \hline \end{array} \qquad \begin{array}{r} 11{,}560 \\ \times \quad .33 \\ \hline \end{array}$$

- Do the calculations.

$$\begin{array}{r} 12{,}290 \\ \times \quad .28 \\ \hline 98320 \\ 24580 \\ \hline 3441.20 \end{array} \qquad \begin{array}{r} 11{,}560 \\ \times \quad .33 \\ \hline 34680 \\ 34680 \\ \hline 3814.80 \end{array}$$

- Check the accuracy of the answers.

$$\begin{array}{r} .28 \\ 12{,}290\overline{)3441.20} \\ 2458\ 0 \\ \hline 983\ 20 \\ 983\ 20 \\ \hline 0 \end{array} \qquad \begin{array}{r} .28 \\ 11.560\overline{)3814.80} \\ 3468\ 0 \\ \hline 346\ 80 \\ 346\ 80 \\ \hline 0 \end{array}$$

- Draw the conclusion.

  More travelers went to the Caribbean and Central America in Year 2 than in Year 1.

CHECK the solution

Make sure the solution solves the work problem:

- Was the defined purpose accomplished?

  Yes, the results of the calculations were analyzed and interpreted. A statement about the year that more people travelled was made.

- Is the solution to the work problem reasonable?

  Since 3814.8 is higher than 3441.2, it is reasonable to conclude that more travelers went to the Caribbean and Central America in Year 2 than in Year 1. Also, since the amounts and percents Al started with were fairly similar, it is reasonable that the two resulting amounts do not differ much.

Problem-Solving Practice Questions

Use the DEFINE, PLAN, SOLVE, and CHECK steps to solve the following problem.

Work Problem

Gary is studying the travel trends for Europe and the Mediterranean. Did more people travel to Europe in Year 3 or Year 4?

DEFINE the problem

- What is the expected outcome? ________________________

- What is the purpose? ________________________

PLAN the solution

- What data are needed? ______________________________
- Where can the data be found? ______________________________
- What is already known? ______________________________
- Which operations should be used? ______________________________

SOLVE the problem

- Select the relevant data. ______________________________
- Set up the calculations.
- Do the calculations.
- Check the accuracy of the answers.

- Draw the conclusion. ______________________________

CHECK the solution

Make sure the solution solves the work problem:

- Was the defined purpose accomplished? ______________________________

- Is the solution to the work problem reasonable? ______________________________

Answers to Problem-Solving Practice questions appear on page 360.

On Your Own

Solve the following work problems. Refer to the table below.

U.S. Travelers to Foreign Countries
(excluding Canada and Mexico)
(in thousands)

	Year 1	Year 2	Year 3	Year 4	Year 5	Year 6
Total Overseas Travelers	12,290	11,560	11,810	12,310	13,310	12,520
Europe & Mediterranean	52%	46%	47%	45%	44%	45%
Caribbean & Cent. Amer.	28%	33%	31%	30%	31%	32%
South America	7%	6%	8%	8%	7%	9%
Asia	6%	8%	6%	9%	10%	8%
Africa	4%	3%	2%	3%	3%	3%
Australia	2%	2%	3%	2%	3%	2%
Other	1%	1%	1%	1%	1%	1%

Work Problem A

Gary is analyzing data about travelers to Europe and the Mediterranean. Did more people travel to Europe and the Mediterranean in Year 4 or Year 5?

Answer: ____________________

Work Problem B

Gary would like to compare the Years 3 through 6 in terms of the number of travelers who went to Europe and the Mediterranean. Rank those years from highest to lowest according to number of travelers.

Years:

1. ____________

2. ____________

3. ____________

4. ____________

Work Problem C

Gary is analyzing data about travel to the Caribbean and Central America. Did more people travel there in Year 3 or Year 4?

Answer: ____________

Work Problem D

Did more people travel to the Caribbean and Central America in Year 4 or Year 5?

Answer: ____________

Work Problem E

Gary would like to compare the numbers of travelers who went to the Caribbean and Central America during the years Year 1 through Year 6. Rank those years from highest to lowest according to numbers of travelers.

Years:

1. ____________ **2.** ____________

3. ____________ **4.** ____________

5. ____________ **6.** ____________

Skills Practice: Comparing Percents

The editorial department spent 95% of its monthly budget. The sales and marketing department spent 80% of its monthly budget. Which department spent more dollars?

Although 95 is greater than 80, the editorial department did not necessarily spend more money. There is not enough information given in the problem to answer the question. The monthly budget amounts for each department must be identified.

If the two departments had the same budget of $300,000, then no calculations are necessary. Simply compare the percents: 95% of $300,000 is more than 80% of $300,000. The editorial department spent more money.

Sales and Marketing

80% of $300,000

Editorial

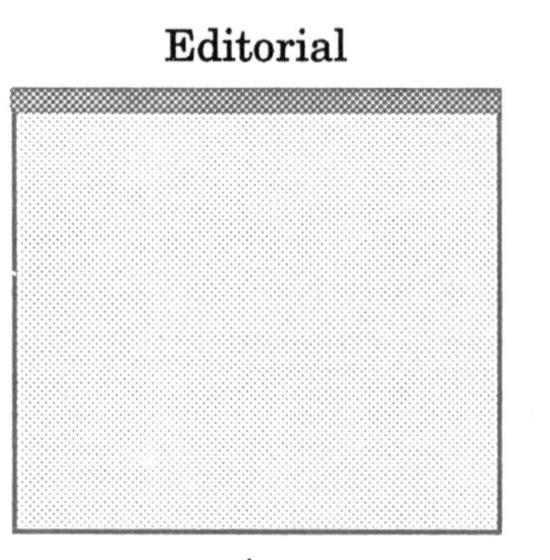

95% of $300,000

However, if the departments' budgets differ, then the percents cannot be compared without calculating first. If the editorial department had a budget of $250,000 and the sales and marketing department had a budget of $350,000, which department spent more money?

Step 1. Calculate the editorial department's spending.

$$\begin{array}{r} \$250{,}000 \\ \underline{\times \quad .95} \\ \$237{,}500 \end{array}$$

Step 2. Calculate the sales and marketing department's spending.

$$\begin{array}{r} \$350{,}000 \\ \underline{\times \quad .80} \\ \$280{,}000 \end{array}$$

Step 3. Compare the amounts.

Since $280,000 is more than $237,000, the marketing department spent more money.

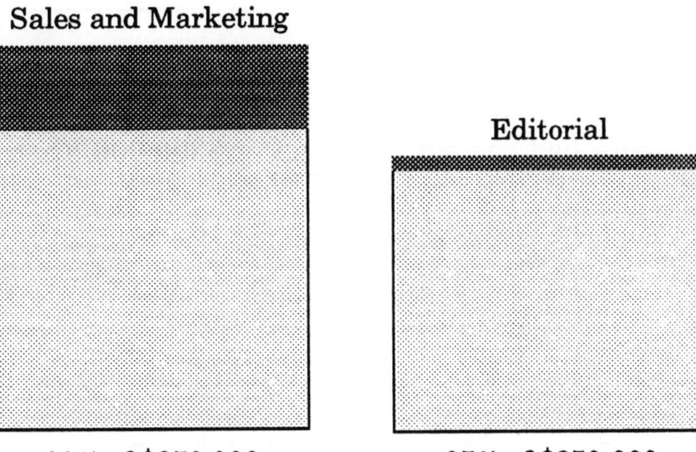

Two percents can be directly compared only when they represent parts of the same number. To compare percents that represent different numbers, first find the amount that each percent represents.

Exercise 1

The table below shows how much money U.S. travelers spent in various regions of the world from Year 1 to Year 5. The percents tell what part of the total for that year was spent in a particular region of the world. For example, travelers spent 30% of $17,560,000,000 in Europe and the Mediterranean in Year 1.

Expenditures of U.S. Travelers in Foreign Countries

(in millions)

	Year 1	Year 2	Year 3	Year 4	Year 5
Expenditures Abroad	$17,560	$18,340	$19,210	$20,430	$20,990
Europe & Mediterranean	30%	31%	29%	29%	30%
Mexico	22%	24%	25%	24%	23%
Canada	18%	16%	15%	17%	18%
Caribbean & Cent. Amer.	11%	10%	11%	9%	9%
Asia	7%	8%	9%	10%	11%
South America	3%	4%	5%	5%	4%
Other	9%	7%	6%	6%	5%

Do the percents described below represent parts of the same number? Write "yes" or "no" on the line.

_____________ **1.** The percent travelers spent in Europe and the Mediterranean in Year 3

The percent travelers spent in the Caribbean and Central America in Year 4

_____________ **2.** The percent travelers spent in Asia in Year 2

The percent travelers spent in Mexico in Year 2

_____________ **3.** The percent travelers spent in Mexico in Year 4

The percent travelers spent in Mexico in Year 5

___________ **4.** The percent travelers spent in South America in Year 3
The percent travelers spent in South America in Year 4

___________ **5.** The percent travelers spent in Asia in Year 3
The percent travelers spent in Canada in Year 4

___________ **6.** The percent travelers spent in Europe and the Mediterranean in Year 5
The percent travelers spent in Mexico in Year 5

___________ **7.** The percent travelers spent in Canada in Year 1
The percent travelers spent in Canada in Year 2

___________ **8.** The percent travelers spent in the Caribbean and Central America in Year 3
The percent travelers spent in the Caribbean and Central America in Year 4

Answers begin on page 361.

Exercise 2

Below are the same percentages as those described in Exercise 1. Compare the two percentages. Put an **X** in front of the percentage that represents the highest dollar amount.

___________ **1.** The percent travelers spent in Europe and the Mediterranean in Year 3
___________ The percent travelers spent in the Caribbean and Central America in Year 4

___________ **2.** The percent travelers spent in Asia in Year 2
___________ The percent travelers spent in Mexico in Year 2

___________ **3.** The percent travelers spent in Mexico in Year 4
___________ The percent travelers spent in Mexico in Year 5

___________ **4.** The percent travelers spent in South America in Year 3
___________ The percent travelers spent in South America in Year 4

__________ **5.** The percent travelers spent in Asia in Year 3
__________ The percent travelers spent in Canada in Year 4

__________ **6.** The percent travelers spent in Europe and the Mediterranean in Year 5
__________ The percent travelers spent in Mexico in Year 5

__________ **7.** The percent travelers spent in Canada in Year 1
__________ The percent travelers spent in Canada in Year 2

__________ **8.** The percent travelers spent in the Caribbean and Central America in Year 3
__________ The percent travelers spent in the Caribbean and Central America in Year 4

Answers begin on page 361.

Exercise 3

Write the phrase "is less than," "is greater than," or "is equal to" on the line to correctly complete each statement.

Example: 10% of 200 ___*is equal to*___ 20% of 100.

1. 5% of 100 __________________ 1% of 500.

2. 20% of 60 __________________ 9% of 60.

3. 50% of 250 __________________ 25% of 500.

4. 45% of 640 __________________ 55% of 600.

5. 8% of 5280 __________________ 10% of 5280.

6. 30% of 600 __________________ 40% of 500.

7. 11% of 1100 __________________ 12% of 950.

8. 80% of 400 __________________ 40% of 800.

Check Yourself

1. In order to directly compare two percents, ______________
   a. the two percents must be equal.
   b. the two percents must represent an equal amount.
   c. the two percents must represent parts of the same amount.
   d. the two percents must have different bases.

2. Which of these statements is true?
   a. 15% of 150 is less than 3% of 150.
   b. 3% of 150 equals 30% of 150.
   c. 30% of 150 is greater than 15% of 150.
   d. Not enough information is provided in statements *a*, *b*, or *c* to determine which one is true.

Refer to the table below to answer questions 3, 4, and 5 and the work problem.

U.S. Travelers to Foreign Countries
(excluding Canada and Mexico)

(in thousands)

	Year 1	Year 2	Year 3	Year 4	Year 5	Year 6
Total Overseas Travelers	12,290	11,560	11,810	12,310	13,310	12,520
Europe & Mediterranean	52%	46%	47%	45%	44%	45%
Caribbean & Cent. Amer.	28%	33%	31%	30%	31%	32%
South America	7%	6%	8%	8%	7%	9%
Asia	6%	8%	6%	9%	10%	8%
Africa	4%	3%	2%	3%	3%	3%
Australia	2%	2%	3%	2%	3%	2%
Other	1%	1%	1%	1%	1%	1%

3. How many people traveled to Asia in Year 2?

a. 924.8
b. 8000
c. 800,000
d. 924,800

4. In Year 2, did more people travel to Asia or South America?

a. Asia
b. South America

5. Did more people travel to Asia in Year 2 or Year 3?

a. Asia
b. South America

Work Problem

Gary is analyzing the data regarding travel to Africa. Did more people travel to Africa in Year 2 or Year 3?

Answer: ____________________

Answers to Problem-Solving Practice Questions

DEFINE the problem

- A statement about which year more people traveled to Europe and the Mediterranean
- To analyze and interpret percentages

PLAN the solution

- How many travelers went to Europe and the Mediterranean during Year 3 and Year 4
- From calculations with the percents on the table
- The percent of travelers who went to Europe and the Mediterranean during Year 3 and Year 4, and the total number of travelers for Year 3 and Year 4
- Multiplication

SOLVE the problem

- The percent of travelers who went to Europe and the Mediterranean during Year 3 and Year 4, and the total number of travelers for Year 3 and Year 4
- $\begin{array}{r} 11{,}810 \\ \times \quad .47 \\ \hline \end{array}$ $\qquad$ $\begin{array}{r} 12{,}310 \\ \times \quad .45 \\ \hline \end{array}$
- 11,810 $\qquad$ 12,310
- $\begin{array}{r} 11{,}810 \\ \times \quad .47 \\ \hline 82670 \\ 47240 \\ \hline 5550.70 \end{array}$ $\qquad$ $\begin{array}{r} 12{,}310 \\ \times \quad .45 \\ \hline 61550 \\ 49240 \\ \hline 5539.50 \end{array}$
- $\begin{array}{r} .47 \\ 11{,}810\overline{)5550.70} \\ \underline{4724\ 0} \\ 826\ 70 \\ \underline{826\ 70} \\ 0 \end{array}$ $\qquad$ $\begin{array}{r} .45 \\ 12{,}310\overline{)5539.50} \\ \underline{4924\ 0} \\ 615\ 50 \\ \underline{615\ 50} \\ 0 \end{array}$
- More travelers went to Europe and the Mediterranean in Year 3 than in Year 4.

CHECK the solution

- The defined purpose was accomplished. The results of the calculations were analyzed and interpreted.
- The solution to the work problem is reasonable since 5550.70 is higher than 5539.50. It is reasonable to conclude that more travelers went to Europe and the Mediterranean in Year 3 than in Year 4.

Answers to Skills Practice Problems

Exercise 1

1. no	**2.** yes	**3.** no	**4.** no
5. no	**6.** yes	**7.** no	**8.** no

Exercise 2

1. first	**2.** second	**3.** first	**4.** second
5. second	**6.** first	**7.** first	**8.** first

Putting It All Together

Work Problem A

The marketing department at Voyager Publishing relies on the table below for information about trends in traveling. What kind of trend is indicated by the number of overseas travelers from Year 2 through Year 5?

Answer: ____________________

U.S. Travelers to Foreign Countries
(excluding Canada and Mexico)

(in thousands)

	Year 1	Year 2	Year 3	Year 4	Year 5	Year 6
Total Overseas Travelers	12,290	11,560	11,810	12,310	13,310	12,520
Europe & Mediterranean	52%	46%	47%	45%	44%	45%
Caribbean & Cent. Amer.	28%	33%	31%	30%	31%	32%
South America	7%	6%	8%	8%	7%	9%
Asia	6%	8%	6%	9%	10%	8%
Africa	4%	3%	2%	3%	3%	3%
Australia	2%	2%	3%	2%	3%	2%
Other	1%	1%	1%	1%	1%	1%

Work Problem B

Gary would like to compare the number of people who went to South America in Year 4 through Year 6. Rank those years from highest to lowest according to number of travelers.

Years:

1. ____________________

2. ____________________

3. ____________________